Algebra through practice

Book 2: Matrices and vector spaces

Algebra through practice

A collection of problems in algebra with solutions

Book 2
Matrices and vector spaces

T.S.BLYTH ○ E.F.ROBERTSON
University of St Andrews

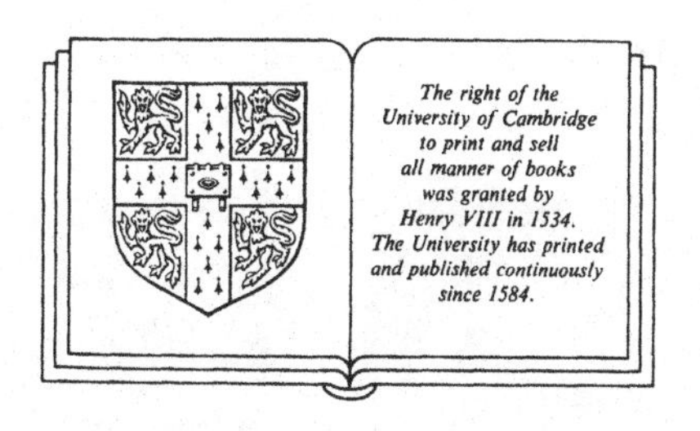
The right of the
University of Cambridge
to print and sell
all manner of books
was granted by
Henry VIII in 1534.
The University has printed
and published continuously
since 1584.

CAMBRIDGE UNIVERSITY PRESS
Cambridge
London New York New Rochelle
Melbourne Sydney

Published by the Press Syndicate of the University of Cambridge
The Pitt Building, Trumpington Street Cambridge CB2 1RP
32 East 57th Street, New York, NY 10022, USA
296 Beaconsfield Parade, Middle Park, Melbourne 3206, Australia

First published 1984

Library of Congress catalogue card number: 83-24013

British Library cataloguing in publication data

Blyth, T. S.
Algebra through practice.
Book 2: Matrices and vector spaces

1. Algebra – Problems, exercises, etc.
I. Title II. Robertson, E. F.
512'.0076 QA157

ISBN 0 521 27286 6

Transferred to digital printing 2003

DJ

Contents

Preface

The aim of this series of problem-solvers is to provide a selection of worked examples in algebra designed to supplement undergraduate algebra courses. We have attempted, mainly with the average student in mind, to produce a varied selection of exercises while incorporating a few of a more challenging nature. Although complete solutions are included, it is intended that these should be consulted by readers only after they have attempted the questions. In this way, it is hoped that the student will gain confidence in his or her approach to the art of problem-solving which, after all, is what mathematics is all about.

The problems, although arranged in chapters, have not been 'graded' within each chapter so that, if readers cannot do problem n this should not discourage them from attempting problem $n+1$. A great many of the ideas involved in these problems have been used in examination papers of one sort or another. Some test papers (without solutions) are included at the end of each book; these contain questions based on the topics covered.

TSB, EFR
St Andrews

Background reference material

Courses on abstract algebra can be very different in style and content. Likewise, textbooks recommended for these courses vary enormously, not only in notation and exposition but also in their level of sophistication. Here is a list of some major texts that are widely used and to which the reader may refer for background material. The subject matter of these texts covers all six books in the *Algebra through practice* series, and in some cases a great deal more. For the convenience of the reader there is given below an indication of which parts of which of these texts is more relevant to the appropriate chapters of this book.

[1] I. T. Adamson, *Introduction to Field Theory*, Cambridge University Press, 1982.

[2] F. Ayres, Jr, *Modern Algebra*, Schaum's Outline Series, McGraw-Hill, 1965.

[3] D. Burton, *A First Course in Rings and Ideals*, Addison-Wesley, 1970.

[4] P. M. Cohn, *Algebra*, Vol. I, Wiley, 1982.

[5] D. T. Finkbeiner II, *Introduction to Matrices and Linear Transformations*, Freeman, 1978.

[6] R. Godement, *Algebra*, Kershaw, 1983.

[7] J. A. Green, *Sets and Groups*, Routledge and Kegan Paul, 1965.

[8] I. N. Herstein, *Topics in Algebra*, Wiley, 1977.

[9] K. Hoffman and R. Kunze, *Linear Algebra*, Prentice Hall, 1971.

[10] S. Lang, *Introduction to Linear Algebra*, Addison-Wesley, 1970.

[11] S. Lipschutz, *Linear Algebra*, Schaum's Outline Series, McGraw-Hill, 1974.

[12] I. D. Macdonald, *The Theory of Groups*, Oxford University Press, 1968.

[13] S. MacLane and G. Birkhoff, *Algebra*, Macmillan, 1968.

[14] N. H. McCoy, *Introduction to Modern Algebra*, Allyn and Bacon, 1975.

[15] J. J. Rotman, *The Theory of Groups: An Introduction*, Allyn and Bacon, 1973.

[16] I. Stewart, *Galois Theory*, Chapman and Hall, 1973.
[17] I. Stewart and D. Tall, *The Foundations of Mathematics*, Oxford University Press, 1977.

References useful for Book 2

1: Matrices and linear equations [2, Chapter 14], [5, Chapters 1, 4], [9, Chapter 1], [10, Chapter 3], [11, Chapters 2, 3].
2: Inverses and determinants [2, Chapter 14], [5, Chapter 5], [9, Chapter 1] [10, Chapter 7], [11, Chapters 3, 8].
3: Eigenvalues and diagonalisation [2, Chapter 15], [11, Chapter 9].
4: Vector spaces [2, Chapter 15], [5, Chapter 2], [9, Chapter 2], [10, Chapter 2], [11, Chapters 4, 5].
5: Linear mappings [5, Chapter 3], [9, Chapter 3], [10, Chapter 4], [11, Chapters 6, 7].
6: Inner product spaces [5, Chapter 8], [9, Chapter 8], [10, Chapter 6], [11, Chapter 13].

In [2] the author uses 'row canonical matrix' for Hermite normal form, and writes linear mappings on the right. In [5] the author uses 'reduced echelon form' for Hermite normal form. This also occurs in [9], and in [11]. The term 'null space' is used in [9] for kernel. Also, some of these texts use 'characteristic vectors' and 'characteristic values' for eigenvectors and eigenvalues.

1: Matrices and linear equations

The basic laws of matrix algebra: associativity of products (when they are defined), distributivity of multiplication over addition (when meaningful), and the like should be well known to the reader. We often use notation such as $\mathrm{Mat}_{m\times n}(\mathbb{R})$ to denote the set of $m \times n$ matrices whose entries belong to the set $\mathbb{R}$ of real numbers. We often write $A \in \mathrm{Mat}_{m\times n}(\mathbb{R})$ in the form $A = [a_{ij}]_{m\times n}$, and the identity $n \times n$ matrix as $I_n = [\delta_{ij}]$ where

$$\delta_{ij} = \begin{cases} 1 & \text{if } i = j; \\ 0 & \text{if } i \neq j. \end{cases}$$

We assume that the reader is familiar with the transpose $A^t = [a_{ji}]_{n\times m}$ of $A = [a_{ij}]_{m\times n}$, with the notions of symmetric ($A^t = A$) and skew-symmetric ($A^t = -A$) matrices and the properties $(A^t)^t = A$, $(A + B)^t = A^t + B^t$, $(AB)^t = B^tA^t$.

We have included in this introductory section a few questions relating matrices to two-dimensional coordinate geometry, which is one of the applications of matrices that is usually covered at this level.

Another basic application is to the solution of systems of linear equations. We assume that the reader has the necessary background knowledge. This includes the reduction of an $m \times n$ matrix to row-echelon form and then to Hermite (normal) form. In particular, the reader should know that the rank of an $m \times n$ matrix is the number of non-zero rows in any row-echelon form, and that a system of equations $A\boldsymbol{x} = \boldsymbol{b}$ has a solution if and only if the rank of the coefficient matrix A is the same as that of the augmented matrix $A|\boldsymbol{b}$.

1.1 Compute the following matrix products:

$$\begin{bmatrix} 1 & 3 & 0 & 4 \\ 2 & 1 & 1 & 0 \end{bmatrix} \begin{bmatrix} 1 & 0 \\ 0 & 1 \\ 1 & 0 \\ 0 & 1 \end{bmatrix}; \quad \begin{bmatrix} 1 & 3 & 2 \\ 2 & 1 & 0 \\ 1 & 0 & 1 \end{bmatrix} \begin{bmatrix} 1 & 2 \\ 0 & 1 \\ 1 & 1 \end{bmatrix};$$

$$\begin{bmatrix} 3 & 1 & -2 \\ 2 & -2 & 0 \\ -1 & 1 & 2 \end{bmatrix} \begin{bmatrix} 1 & 1 & 1 \\ 1 & -1 & 1 \\ 0 & 1 & 2 \end{bmatrix}; \quad \begin{bmatrix} 1 \\ 2 \\ 3 \\ 4 \end{bmatrix} [1 \quad 2 \quad 3 \quad 4]$$

$$[1 \quad 2 \quad 3 \quad 4] \begin{bmatrix} 1 \\ 2 \\ 3 \\ 4 \end{bmatrix}.$$

1.2 Compute the matrix product

$$[x \quad y \quad 1] \begin{bmatrix} a & h & g \\ h & b & f \\ g & f & c \end{bmatrix} \begin{bmatrix} x \\ y \\ 1 \end{bmatrix}.$$

Hence express in matrix notation the equations

(*a*) $x^2 + 9xy + y^2 + 8x + 5y + 2 = 0$;

(*b*) $\dfrac{x^2}{\alpha^2} + \dfrac{y^2}{\beta^2} = 1$;

(*c*) $xy = \alpha^2$;

(*d*) $y^2 = 4\alpha x$.

1.3 If $A = \begin{bmatrix} 0 & 1 \\ 0 & 1 \end{bmatrix}$ and $B = \begin{bmatrix} -1 & -1 \\ 0 & 0 \end{bmatrix}$ prove that

$$(A + B)^2 \neq A^2 + 2AB + B^2$$

but that

$$(A + B)^3 = A^3 + 3A^2B + 3AB^2 + B^3.$$

1.4 Let A be the matrix

$$\begin{bmatrix} 0 & a & a^2 & a^3 \\ 0 & 0 & a & a^2 \\ 0 & 0 & 0 & a \\ 0 & 0 & 0 & 0 \end{bmatrix}.$$

Define the matrix B by

$$B = A - \tfrac{1}{2}A^2 + \tfrac{1}{3}A^3 - \tfrac{1}{4}A^4 + \cdots$$

Show that this series has only finitely many terms different from zero and calculate B. Show also that the series

$$B + \frac{1}{2!}B^2 + \frac{1}{3!}B^3 + \cdots$$

has only a finite number of non-zero terms and that its sum is A.

1.5 Find all $X \in \mathrm{Mat}_{2\times 2}(\mathbb{R})$ such that $X^2 = I_2$.

1.6 Find all $A \in \mathrm{Mat}_{2\times 2}(\mathbb{C})$ such that $A^2 = -I_2$. Show that there are no 2×2 real diagonal matrices A with $A^2 = -I_2$, but that there are infinitely many 2×2 real matrices A with $A^2 = -I_2$. Deduce that for every even positive integer n there are infinitely many $n \times n$ real matrices A with $A^2 = -I_n$. Is this so for every odd positive integer n?

1.7 Show that every $A \in \mathrm{Mat}_{2\times 2}(\mathbb{C})$ which is such that $A^2 = 0$ may be written in the form

$$\begin{bmatrix} ab & a^2 \\ -b^2 & -ab \end{bmatrix}$$

for some $a, b \in \mathbb{C}$. Is it true that every $A \in \mathrm{Mat}_{2\times 2}(\mathbb{R})$ such that $A^2 = 0$ is of this form with $a, b \in \mathbb{R}$?

1.8 Show that a real 2×2 matrix $\begin{bmatrix} a & b \\ c & d \end{bmatrix}$ may be expressed as a product $\begin{bmatrix} x \\ y \end{bmatrix} [u \quad v]$ for some $x, y, u, v \in \mathbb{R}$ if, and only if, $ad = bc$.

1.9 For $A, B \in \mathrm{Mat}_{n\times n}(\mathbb{R})$ define $[AB] = AB - BA$.

(*a*) Prove that the following identities hold:

(i) $[[AB]C] + [[BC]A] + [[CA]B] = 0$;

(ii) $[(A + B)C] = [AC] + [BC]$;

(iii) $[[[AB]C]D] + [[[BC]D]A] + [[[CD]A]B] + [[[DA]B]C] = 0$.

(*b*) Show by means of an example that in general

$$[[AB]C] \neq [A[BC]].$$

1.10 Consider the complex 2×2 matrices

$$X = \begin{bmatrix} \frac{1}{2}i & 0 \\ 0 & -\frac{1}{2}i \end{bmatrix}, \quad Y = \begin{bmatrix} 0 & \frac{1}{2} \\ -\frac{1}{2} & 0 \end{bmatrix}, \quad Z = \begin{bmatrix} 0 & \frac{1}{2}i \\ \frac{1}{2}i & 0 \end{bmatrix}.$$

(*a*) Show that $AB = -BA$ for all $A, B \in X, Y, Z$ with $A \neq B$.

(*b*) Compute $AB - BA$ for each distinct pair $A, B \in \{X, Y, Z\}$ and comment on your answer.

(*c*) Prove that the 3×3 real matrices

$$X' = \begin{bmatrix} 0 & 0 & 0 \\ 0 & 0 & -1 \\ 0 & 1 & 0 \end{bmatrix}, \quad Y' = \begin{bmatrix} 0 & 0 & 1 \\ 0 & 0 & 0 \\ -1 & 0 & 0 \end{bmatrix}, \quad Z' = \begin{bmatrix} 0 & -1 & 0 \\ 1 & 0 & 0 \\ 0 & 0 & 0 \end{bmatrix}$$

satisfy the same properties as X, Y, Z.

1.11 (*a*) Show that if A and B are 2×2 matrices then the sum of the diagonal elements of $AB - BA$ is zero.

(*b*) If E is a 2×2 matrix and the sum of the diagonal elements of E is zero show that $E^2 = \lambda I_2$ for some scalar λ.

(*c*) Deduce from (*a*) and (*b*) that if A, B, C are 2×2 matrices then

$$(AB - BA)^2 C = C(AB - BA)^2.$$

1.12 Let A, B be $n \times n$ matrices with A symmetric and B skew-symmetric. Determine which of the following are symmetric and which are skew-symmetric:

$$AB + BA; \quad AB - BA; \quad A^2; \quad B^2; \quad A^p B^q A^p.$$

1.13 Let $\boldsymbol{x}$ and $\boldsymbol{y}$ be $n \times 1$ matrices. Show that the matrix $A = \boldsymbol{x}\boldsymbol{y}^t - \boldsymbol{y}\boldsymbol{x}^t$ is of size $n \times n$ and is skew-symmetric. Show also that $\boldsymbol{x}^t\boldsymbol{y}$ and $\boldsymbol{y}^t\boldsymbol{x}$ are of size 1×1 and are equal.

If $\boldsymbol{x}^t\boldsymbol{x} = \boldsymbol{y}^t\boldsymbol{y} = [1]$ and $\boldsymbol{x}^t\boldsymbol{y} = \boldsymbol{y}^t\boldsymbol{x} = [k]$, prove that $A^3 = (k^2 - 1)A$.

1.14 If A is a square matrix such that $A^2 = A$ and $(A - A^t)^2 = 0$ prove that $(AA^t)^2 = AA^t$.

1.15 Suppose that in the cartesian plane the coordinate axes are rotated in an anticlockwise direction through an angle ϑ. Show that the 'new' coordinates (x', y') of the point P whose 'old' coordinates are (x, y) are given by

$$\begin{bmatrix} x' \\ y' \end{bmatrix} = R_\vartheta \begin{bmatrix} x \\ y \end{bmatrix}$$

where R_ϑ is the rotation matrix

$$\begin{bmatrix} \cos\vartheta & \sin\vartheta \\ -\sin\vartheta & \cos\vartheta \end{bmatrix}.$$

Prove that, for rotations ϑ and φ,

$$R_\vartheta R_\varphi = R_{\vartheta+\varphi} = R_\varphi R_\vartheta.$$

The hyperbola $x^2 - y^2 = 1$ is rotated anti-clockwise about the origin through 45°; find its new equation.

1.16 Two similar sheets of graph paper are pinned together at the origin and the sheets are rotated. If the point $(1, 0)$ of the top sheet lies directly above the point $(\frac{5}{13}, \frac{12}{13})$ of the bottom sheet, above what point of the bottom sheet does the point $(2, 3)$ of the top sheet lie?

1.17 For every point (x, y) of the cartesian plane let (x', y') be its reflection in the x-axis. Find the matrix M such that

$$\begin{bmatrix} x' \\ y' \end{bmatrix} = M \begin{bmatrix} x \\ y \end{bmatrix}.$$

1.18 In the cartesian plane let L be a line passing through the origin and making an angle ϑ with the x-axis. For every point (x, y) of the plane let (x_L, y_L) be its reflection in the line L. Prove that

$$\begin{bmatrix} x_L \\ y_L \end{bmatrix} = \begin{bmatrix} \cos 2\vartheta & \sin 2\vartheta \\ \sin 2\vartheta & -\cos 2\vartheta \end{bmatrix} \begin{bmatrix} x \\ y \end{bmatrix}.$$

1.19 In the cartesian plane let L be a line passing through the origin and making an angle ϑ with the x-axis. For every point (x, y) of the plane let (x^*, y^*) be the foot of the perpendicular from (x, y) onto L. Prove that

$$\begin{bmatrix} x^* \\ y^* \end{bmatrix} = \begin{bmatrix} \cos^2\vartheta & \sin\vartheta\cos\vartheta \\ \sin\vartheta\cos\vartheta & \sin^2\vartheta \end{bmatrix} \begin{bmatrix} x \\ y \end{bmatrix}.$$

1.20 Find the Hermite normal form of each of the following matrices:

$$\begin{bmatrix} 1 & 2 & 3 \\ 3 & 1 & 2 \\ 5 & 5 & 8 \end{bmatrix}; \quad \begin{bmatrix} 1 & 2 & 3 \\ 3 & 1 & 2 \\ 2 & 3 & 1 \end{bmatrix}; \quad \begin{bmatrix} 2 & 1 & 3 \\ 3 & 1 & 3 \\ 0 & 1 & 3 \end{bmatrix}.$$

1.21 Reduce to row-echelon form the augmented matrix of the system of equations

$$\begin{aligned} x + 2y \phantom{{}+3z} + 3t &= 1 \\ x + 2y + 3z + 3t &= 3 \\ x \phantom{{}+2y} + z + t &= 3 \\ x + y + z + 2t &= 1. \end{aligned}$$

Deduce that the system has no solution.

1.22 For what value of λ does the system of equations

$$\begin{aligned} x + y \phantom{{}+z} + t &= 4 \\ 2x \phantom{{}+y+z} - 4t &= 7 \\ x + y + z &= 5 \\ x - 3y - z - 10t &= \lambda \end{aligned}$$

have a solution? Find the general solution when λ takes this value.

1.23 What conditions must the integers a, b, c satisfy in order that the system of equations

$$\begin{aligned} 2w - x + y - 3z &= a \\ w + x - y &= b \\ 4w + x - y - 3z &= c \end{aligned}$$

has integer solutions?

1.24 If $a, b, c, d \in \mathbb{R}$ are all greater than 0 prove that the system of equations

$$\begin{aligned} x + y + z + t &= a \\ x - y - z + t &= b \\ -x - y + z + t &= c \\ -3x + y - 3z - 7t &= d \end{aligned}$$

has no solutions.

1.25 Show that the equations

$$\begin{aligned} 2x + y + z &= -6\beta \\ \gamma x + 3y + 2z &= 2\beta \\ 2x + y + (\gamma + 1)z &= 4 \end{aligned}$$

have a unique solution except when $\gamma = 0$ and when $\gamma = 6$. If $\gamma = 0$ prove that there is only one value of β for which a solution exists and find the general solution in this case. Discuss the situation when $\gamma = 6$.

1.26 Show that the equations

$$\begin{aligned} x - y \qquad\quad - u - 5t &= \alpha \\ 2x + y - z - 4u + t &= \beta \\ x + y + z - 4u - 6t &= \gamma \\ x + 4y + 2z - 8u - 5t &= \delta \end{aligned}$$

have a solution if and only if

$$8\alpha - \beta - 11\gamma + 5\delta = 0.$$

Find the general solution when $\alpha = \beta = -1, \gamma = 3, \delta = 8$

2: Inverses and determinants

An $n \times n$ matrix M has an inverse if there is a (necessarily unique) $n \times n$ matrix M^{-1} such that $MM^{-1} = I_n = M^{-1}M$. (*Note.* It can in fact be shown that one of these equations is enough; if $MX = I_n$ then $XM = I_n$.) Inverses have the properties that $(M^{-1})^{-1} = M$ and $(AB)^{-1} = B^{-1}A^{-1}$. Elementary row operations may be used in deciding whether or not M is invertible. Briefly, to apply an elementary row operation to M is equivalent to pre-multiplying M by an elementary matrix E (namely, that which is obtained by applying the row operation in question to I_n). Starting with an array $M|I_n$ and applying a sequence of row operations we obtain a sequence

$$M|I_n \to E_1M|E_1 \to E_2E_1M|E_2E_1 \to \cdots \to EM|E$$

where $E = E_nE_{n-1} \cdots E_2E_1$. Choosing $E_1, \ldots, E_n$ such that EM is the Hermite form of M, we have that $EM = I_n$ if and only if $E = M^{-1}$. Thus, if the Hermite form of M is I_n then M is invertible, and its inverse is obtained by applying to I_n the same sequence of row operations as was applied to M. If the Hermite form of M is not I_n then M is not invertible.

The determinant of an $n \times n$ matrix M is denoted by $\det M$ or $|M|$. The determinant of a 2×2 matrix

$$A = \begin{bmatrix} a_{11} & a_{12} \\ a_{21} & a_{22} \end{bmatrix}$$

is the cross-product $|A| = a_{11}a_{22} - a_{21}a_{12}$. The determinant of an $n \times n$ matrix is defined inductively as

$$\sum_{\sigma} \epsilon_\sigma a_{1,\sigma(1)} a_{2,\sigma(2)} \cdots a_{n,\sigma(n)}$$

where the sum is over all permutations σ of $\{1, \ldots, n\}$, and ϵ_σ is 1 if σ is a even permutation and -1 if σ is an odd permutation. This definition (and what it involves in general) will not be required. What is assumed is that the reader has (at least on trust!) a knowledge of the Laplace expansion method of computing $\det M$, namely that if M_{ij} is the matrix obtained from M by deleting the ith row and the jth column then

$$\det M = \sum_{j=1}^{n} (-1)^{i+j} a_{ij} \det M_{ij}.$$

This is the Laplace expansion of $\det M$ via the ith row; it can be shown that it is independent of the choice of i. A similar result holds for columns. Elementary row and column operations are used to create zero entries often before applying a Laplace expansion, in evaluating determinants. Fundamental properties of determinants are that $\det M = \det M^t$, $\det AB = \det A \,.\det B$ and, if M is invertible, $\det M^{-1} = (\det M)^{-1}$.

2.1 Find the inverses of the following matrices:

$$\begin{bmatrix} 3 & 1 & 2 \\ 1 & 2 & 1 \\ 1 & 1 & 1 \end{bmatrix}; \quad \begin{bmatrix} 5 & 3 & 2 \\ 2 & 3 & 1 \\ 7 & 5 & 3 \end{bmatrix}; \quad \begin{bmatrix} 1 & 0 & 0 \\ 1 & 2 & 0 \\ 1 & 2 & 3 \end{bmatrix}.$$

2.2 Determine the values of the real number x for which the matrix

$$\begin{bmatrix} x & 2 & 0 & 3 \\ 1 & 2 & 3 & 3 \\ 1 & 0 & 1 & 1 \\ 1 & 1 & 1 & 3 \end{bmatrix}$$

is invertible.

2.3 Let $A \in \mathrm{Mat}_{6\times 6}(\mathrm{IR})$ be given by

$$A = \begin{bmatrix} 0 & a & 0 & 0 & 0 & 0 \\ f & 0 & b & 0 & 0 & 0 \\ 0 & g & 0 & c & 0 & 0 \\ 0 & 0 & h & 0 & d & 0 \\ 0 & 0 & 0 & k & 0 & e \\ 0 & 0 & 0 & 0 & m & 0 \end{bmatrix}.$$

Prove that A is invertible if and only if a, c, e, f, h, m are non-zero, and find A^{-1} in this case.

2.4 Given the matrices

$$A = \begin{bmatrix} b+8c & 2c-2b & 4b-4c \\ 4c-4a & c+8b & 2a-2c \\ 2b-2a & 4a-4b & a+8b \end{bmatrix}, \quad P = \begin{bmatrix} 0 & 1 & 2 \\ 2 & 0 & 1 \\ 1 & 2 & 0 \end{bmatrix},$$

find P^{-1} and evaluate $P^{-1}AP$. Hence evaluate $\det A$.

2.5 If $\Delta \in \mathrm{Mat}_{4\times 4}(\mathrm{IR})$ is given by

$$\Delta = \begin{bmatrix} 1 & 1 & 1 & 1 \\ a & x & b & c \\ a^2 & x^2 & b^2 & c^2 \\ a^3 & x^3 & b^3 & c^3 \end{bmatrix}$$

express $\det \Delta$ as a product of factors.

2.6 Solve the equation

$$\det \begin{bmatrix} x & a & a & a \\ a & x & a & a \\ a & a & x & a \\ a & a & a & x \end{bmatrix} = 0.$$

2.7 Consider the 4×4 real matrix

$$\begin{bmatrix} x & a & ? & ? \\ y^2 & y & a & ? \\ yz^2 & z^2 & z & a \\ yzt^2 & zt^2 & t^2 & t \end{bmatrix}$$

Show that, whatever the entries marked ? may be, this matrix has determinant $(x-ay)(y-az)(z-at)t$.

2.8 Let $B_r = [\beta_{ij}] \in \mathrm{Mat}_{r\times r}(\mathrm{IR})$ be defined by

$$\beta_{ij} = \begin{cases} b & \text{if } i \leqslant j; \\ a & \text{if } i = j+1; \\ -b & \text{if } i > j+1. \end{cases}$$

Prove that $\det B_{n-1} = (-1)^n b(a-b)^{n-2}$.

Hence show that if $A_r = [\alpha_{ij}] \in \mathrm{Mat}_{r\times r}(\mathrm{IR})$ is given by

$$\alpha_{ij} = \begin{cases} b & \text{if } i<j; \\ a & \text{if } i=j; \\ -b & \text{if } i>j \end{cases}$$

then $\det A_n = (a+b)\det A_{n-1} - b(a-b)^{n-1}$.

Deduce that

$$\det A_n = \tfrac{1}{2}\{(a+b)^n + (a-b)^n\}.$$

2.9 Let $A_n = [\alpha_{ij}] \in \mathrm{Mat}_{n\times n}(\mathrm{IR})$ be given by

$$\alpha_{ij} = \begin{cases} 0 & \text{if } |i-j|>1; \\ 1 & \text{if } |i-j|=1; \\ 2\cos\vartheta & \text{if } i=j. \end{cases}$$

If $\Delta_n = \det A_n$, prove that $\Delta_{n+2} - 2\cos\vartheta\,\Delta_{n+1} + \Delta_n = 0$. Hence show by induction that, for $0<\vartheta<\pi$,

$$\Delta_n = \frac{\sin(n+1)\vartheta}{\sin\vartheta}.$$

2.10 Let $A_n = [\alpha_{ij}] \in \mathrm{Mat}_{n\times n}(\mathrm{IR})$ be given by

$$\alpha_{ij} = \begin{cases} b_i & \text{if } i\neq j; \\ a_i+b_i & \text{if } i=j<n; \\ b_n & \text{if } i=j=n. \end{cases}$$

Prove that $\det A_n = b_n \prod_{i=1}^{n-1} a_i$.

If $B_n = [\beta_{ij}] \in \mathrm{Mat}_{n\times n}(\mathrm{IR})$ is given by

$$\beta_{ij} = \begin{cases} b_i & \text{if } i\neq j; \\ a_i+b_i & \text{if } i=j, \end{cases}$$

prove that $\det B_n = \det A_n + a_n \det B_{n-1}$. Hence deduce that

$$\det B_n = \prod_{i=1}^{n} a_i + \sum_{i=1}^{n}\left(b_i \prod_{j\neq i} a_j\right).$$

2.11 A real matrix $A = [a_{ij}]_{n\times n}$ is said to be dominated by its diagonal elements if, for $1 \leqslant r \leqslant n$,

$$|a_{rr}| > \sum_{i\neq r} |a_{ri}|.$$

Prove that if A is dominated by its diagonal elements then $\det A \neq 0$.

2.12 If A and B are square matrices of the same order prove that

$$\det\begin{bmatrix} A & B \\ B & A \end{bmatrix} = \det(A+B)\det(A-B).$$

2.13 Let $M = \begin{bmatrix} P & Q \\ R & S \end{bmatrix}$ where P, Q, R, S are square matrices of the same order with P non-singular. Find a matrix N of the form $\begin{bmatrix} A & 0 \\ B & C \end{bmatrix}$ such that

$$NM = \begin{bmatrix} I & P^{-1}Q \\ 0 & S - RP^{-1}Q \end{bmatrix}.$$

Hence show that if $PR = RP$ then $\det M = \det(PS - RQ)$, and that if $PQ = QP$ then $\det M = \det(SP - RQ)$.

3: Eigenvalues and diagonalisation

In this chapter the emphasis is on the notions of eigenvalue and eigenvector of a square matrix A. These are respectively a scalar λ and a non-zero column matrix $\boldsymbol{x}$ such that $A\boldsymbol{x} = \lambda\boldsymbol{x}$. In order to compute the λ and the $\boldsymbol{x}$ one begins by considering the system of equations $(A - \lambda I_n)\boldsymbol{x} = 0$. These have a non-zero solution if and only if $\det(A - \lambda I_n) = 0$. The left hand side of this equation is a polynomial of degree n which, when made monic, becomes the characteristic polynomial $\chi_A(X)$ of A. The zeros of the characteristic polynomial are thus the eigenvalues. It can be shown that every $n \times n$ matrix A satisfies its characteristic polynomial (the Cayley-Hamilton theorem). The minimum polynomial $m_A(X)$ of A is the monic polynomial of least degree satisfied by A. It has degree less than or equal to that of $\chi_A(X)$ and divides $\chi_A(X)$.

If the $n \times n$ matrix A has n distinct eigenvalues $\lambda_1, \ldots, \lambda_n$ and if $\boldsymbol{x}_1, \ldots, \boldsymbol{x}_n$ are corresponding eigenvectors then the matrix P whose ith column is $\boldsymbol{x}_i$ for each i is such that P^{-1} exists and $P^{-1}AP = D$ where $D = [d_{ij}] = \operatorname{diag}\{\lambda_1, \ldots, \lambda_n\}$ is the diagonal matrix with $d_{ii} = \lambda_i$ for every i. When A is real and symmetric, P can be chosen to be orthogonal ($P^{-1} = P^t$); this is achieved by normalising the eigenvectors. The same procedure may be applied when A has less than n distinct eigenvalues provided that one can find n linearly independent eigenvectors.

3.1 Determine the characteristic polynomial and the minimum polynomial of each of the following matrices (A, B, C, D)

$$\begin{bmatrix} 1 & -4 & 0 \\ 2 & -2 & -2 \\ -2.5 & 1 & -2 \end{bmatrix}; \quad \begin{bmatrix} 1 & 2 & 3 \\ 0 & 1 & 2 \\ 0 & 0 & 1 \end{bmatrix}; \quad \begin{bmatrix} 1 & 1 & 0 \\ -1 & 1 & 1 \\ 0 & 1 & -1 \end{bmatrix}; \quad \begin{bmatrix} 0 & 0 & 2 \\ 1 & 0 & -1 \\ 0 & 1 & 1 \end{bmatrix}$$

3.2 If $A, B \in \mathrm{Mat}_{n\times n}(\mathrm{IR})$ are such that $I_n - AB$ is invertible, prove that so also is $I_n - BA$.

Deduce that, for all $A, B \in \mathrm{Mat}_{n\times n}(\mathrm{IR})$, AB and BA have the same eigenvalues.

3.3 Let A, B be square matrices over $\mathbb{C}$ and suppose that there exist rectangular matrices P, Q over $\mathbb{C}$ such that $A = PQ$ and $B = QP$. If $h(X)$ is any polynomial with complex coefficients prove that $A\,h(A) = P\,h(B)\,Q$.

Hence show that, if $m_A(X)$, $m_B(X)$ are the minimum polynomials of A, B respectively, then $Am_B(A) = 0 = Bm_A(B)$. Deduce that one of the following holds:

$$m_A(X) = m_B(X); \quad m_A(X) = Xm_B(X); \quad m_B(X) = Xm_A(X).$$

Express the $r \times r$ matrix

$$\begin{bmatrix} 1 & 1 & \dots & 1 \\ 2 & 2 & \dots & 2 \\ \vdots & \vdots & & \vdots \\ r & r & \dots & r \end{bmatrix}$$

as the product of a column matrix and a row matrix. Hence find its minimum polynomial.

3.4 Given the real $n \times n$ matrix $A = [a_{ij}]$ let $k = \max |a_{ij}|$. Prove that, for all positive integers r,

$$|[A^r]_{ij}| \leqslant k^r n^{r-1}.$$

For every scalar β, associate with A the infinite series

$$S_\beta(A) \equiv I_n + \beta A + \beta^2 A^2 + \dots + \beta^r A^r + \dots.$$

We say that $S_\beta(A)$ converges if each of the series

$$\delta_{ij} + \beta[A]_{ij} + \beta^2[A^2]_{ij} + \dots + \beta^r[A^r]_{ij} + \dots$$

converges. Prove that

(*a*) $S_\beta(A)$ converges if $|\beta| < 1/nk$;

(*b*) if $S_\beta(A)$ converges then $I_n - \beta A$ has an inverse which is the sum of the series.

Deduce that if A is a real $n \times n$ matrix and λ is an eigenvalue of A the $|\lambda| \leqslant n \,.\max |a_{ij}|$.

3.5 Show that the matrix

$$\begin{bmatrix} -2 & -3 & -3 \\ -1 & 0 & -1 \\ 5 & 5 & 6 \end{bmatrix}$$

has only two distinct eigenvalues. Find corresponding eigenvectors.

3.6 Find the eigenvalues, and one eigenvector for each eigenvalue, of each of the following matrices:

$$\begin{bmatrix} 1 & 0 & 1 \\ 0 & 1 & 0 \\ 1 & 0 & 1 \end{bmatrix}; \quad \begin{bmatrix} -2 & 5 & 7 \\ 1 & 0 & -1 \\ -1 & 1 & 2 \end{bmatrix}.$$

3.7 For each of the matrices

$$\begin{bmatrix} -3 & -7 & 19 \\ -2 & -1 & 8 \\ -2 & -3 & 10 \end{bmatrix}, \quad \begin{bmatrix} -4 & 0 & -3 \\ 1 & 3 & 1 \\ 4 & -2 & 2 \end{bmatrix}$$

find a matrix T such that $T^{-1}AT$ is diagonal.

3.8 Show that $\begin{bmatrix} 1 \\ i \end{bmatrix}$ and $\begin{bmatrix} i \\ 1 \end{bmatrix}$ are eigenvectors of

$$A = \begin{bmatrix} \cos\vartheta & \sin\vartheta \\ -\sin\vartheta & \cos\vartheta \end{bmatrix}.$$

If $P = \begin{bmatrix} 1 & i \\ i & 1 \end{bmatrix}$ compute the product $P^{-1}AP$.

3.9 Let $A = \begin{bmatrix} a & b \\ c & d \end{bmatrix}$ where $a, b, c, d > 0$. Prove that

(*a*) the eigenvalues of A are real and distinct;

(*b*) A has at least one positive eigenvalue;

(*c*) corresponding to the largest eigenvalue of A there are infinitely many eigenvectors with both components positive.

3.10 For each of the following matrices determine an orthogonal matrix P such that P^tAP is diagonal with the diagonal entries in increasing order of magnitude:

$$\begin{bmatrix} 0 & 1 & 0 \\ 1 & 0 & 1 \\ 0 & 1 & 0 \end{bmatrix}; \quad \begin{bmatrix} 3 & -1 & 0 \\ -1 & 3 & 0 \\ 0 & 0 & 1 \end{bmatrix}.$$

3.11 Suppose that the real matrix $A = \begin{bmatrix} \alpha & \beta \\ 1 & 0 \end{bmatrix}$ has distinct eigenvalues λ_1, λ_2. If $P = \begin{bmatrix} \lambda_1 & \lambda_2 \\ 1 & 1 \end{bmatrix}$ prove that $P^{-1}AP = D = \text{diag}\,\{\lambda_1, \lambda_2\}$. Deduce that, for every positive integer r, $A^r = PD^rP^{-1}$.

Consider the system of recurrence relations defined by

$$u_{n+1} = \alpha u_n + \beta v_n,$$
$$v_{n+1} = u_n,$$

for $n = 0, 1, 2, \ldots$. Using the matrices

$$U_n = \begin{bmatrix} u_n \\ v_n \end{bmatrix} \quad \text{and} \quad A = \begin{bmatrix} \alpha & \beta \\ 1 & 0 \end{bmatrix}$$

express these relations in matrix form and deduce that $U_n = A^{n-1}U_1$. Hence show that

$$u_n = \frac{\lambda_1^n}{\lambda_1 - \lambda_2}(u_1 - \lambda_2 u_0) + \frac{\lambda_2^n}{\lambda_1 - \lambda_2}(\lambda_1 u_0 - u_1).$$

3.12 Let T_n be the tridiagonal $n \times n$ matrix

$$T_n = \begin{bmatrix} 1 & -4 & 0 & \ldots & 0 & 0 \\ 5 & 1 & -4 & \ldots & 0 & 0 \\ 0 & 5 & 1 & \ldots & 0 & 0 \\ \vdots & \vdots & \vdots & & \vdots & \vdots \\ 0 & 0 & 0 & \ldots & 1 & -4 \\ 0 & 0 & 0 & \ldots & 5 & 1 \end{bmatrix}$$

and let $\Delta_n = \det T_n$. For $n \geqslant 2$ let $r_n = \begin{bmatrix} \Delta_n \\ \Delta_{n-1} \end{bmatrix}$. Show that $r_n = Ar_{n-1}$ where $A = \begin{bmatrix} 1 & 20 \\ 1 & 0 \end{bmatrix}$ and deduce that $r_n = A^{n-2}r_2$. By diagonalising A, find A^{n-2} and hence show that

$$\Delta_n = \tfrac{1}{9}(5^{n+1} - (-4)^{n+1}).$$

3.13 For every positive integer n, determine the nth power of the matrix

$$A = \begin{bmatrix} 2 & 2 & 0 \\ 1 & 2 & 1 \\ 1 & 2 & 1 \end{bmatrix}.$$

3.14 Solve the system of equations

$$x_{n+1} = 2x_n + 6y_n$$
$$y_{n+1} = 6x_n - 3y_n$$

given that $x_1 = 0$ and $y_1 = -1$.

3.15 Given the matrix

$$A = \begin{bmatrix} 6.5 & -2.5 & 2.5 \\ -2.5 & 6.5 & -2.5 \\ 0 & 0 & 4 \end{bmatrix}$$

find a matrix B such that $B^2 = A$.

4: Vector spaces

We consider vector spaces V over only the fields $\mathbb{Q}$, $\mathbb{R}$, $\mathbb{C}$ and $\mathbb{Z}_p$. The subspace spanned (or generated) by the set $\{v_1, \ldots, v_n\}$ is the set of linear combinations of $v_1, \ldots, v_n$ and is written span $\{v_1, \ldots, v_n\}$ or $\langle v_1, \ldots, v_n\rangle$ A linearly independent spanning set is called a basis. The number of elements in a basis is independent of the basis and is the dimension dim V of V. If, for example, the ground field is $\mathbb{R}$ and V is of dimension n then V can be regarded as the set $\mathbb{R}^n$ of n-tuples of real numbers. Every linearly independent subset of V can be extended to a basis. A basis is therefore a maximal independent subset; it can also be described as a minimal generating set. If W is a subspace of V then dim $W \leqslant$ dim V with equality if and only if $W = V$.

4.1 For $i = 1, \ldots, n$ let $a_i = (a_{i1}, a_{i2}, \ldots, a_{in})$. Prove that $\{a_1, \ldots, a_n\}$ is a basis of $\mathbb{R}^n$ if and only if $A = [a_{ij}]_{n \times n}$ is invertible.

4.2 Determine which of the following are bases for $\mathbb{R}^3$:

(*a*) $\{(1, 1, 1), (1, 2, 3), (2, -1, 1)\}$;

(*b*) $\{(1, 1, 2), (1, 2, 5), (5, 3, 4)\}$.

4.3 Show that $\{(1, 1, 0, 0), (-1, -1, 1, 2), (1, -1, 1, 3), (0, 1, -1, -3)\}$ is a basis of $\mathbb{R}^4$ and find the coordinates of the vector (a, b, c, d) relative to this basis.

4.4 Extend the linearly independent set $\{(1, -1, 1, -1), (1, 1, -1, 1)\}$ to a basis for $\mathbb{R}^4$.

4.5 Which of the following statements are true? Give a proof for those which are true and a counter-example for those which are false.

(*a*) $\{(x_1, x_2) \in \mathbb{R}^2 \mid x_1 < x_2\}$ is a subspace of $\mathbb{R}^2$.

(*b*) If $\{a_1, a_2, a_3\}$ is a basis for $\mathbb{R}^3$ and b is a non-zero vector in $\mathbb{R}^3$ then $\{b + a_1, a_2, a_3\}$ is also a basis of $\mathbb{R}^3$.

(c) If $\{x_1, \ldots, x_r\}$ is a linearly dependent set of vectors in $\mathbb{R}^n$ then $r < n$.
(d) If $\{x_1, \ldots, x_r\}$ is a spanning set for $\mathbb{R}^n$ then $r \geqslant n$.
(e) The subspace $\{(x, x, x) \mid x \in \mathbb{R}\}$ of $\mathbb{R}^3$ has dimension 3.
(f) The subspace of $\mathbb{R}^3$ spanned by $\{(1, 2, 1), (2, 2, 1)\}$ is $\{(x + y, 2y, y) \mid x, y \in \mathbb{R}\}$.
(g) The subspace of $\mathbb{R}^3$ spanned by $\{(1, 2, 1), (2, 2, 1)\}$ is $\{(2x, 2x + 2y, x + y) \mid x, y \in \mathbb{R}\}$.
(h) If P, Q are subspaces of a finite-dimensional vector space then we have that $P \subseteq Q$ implies $\dim P \leqslant \dim Q$.
(i) If P, Q are subspaces of a finite-dimensional vector space then we have that $\dim P \leqslant \dim Q$ implies $P \subseteq Q$.
(j) The only n-dimensional subspace of $\mathbb{R}^n$ is $\mathbb{R}^n$ itself.

4.6 Determine whether or not the following subsets of $\mathbb{R}^4$ are subspaces:
(a) $U = \{(a, b, c, d) \mid a + b = c + d\}$;
(b) $U = \{(a, b, c, d) \mid a + b = 1\}$;
(c) $U = \{(a, b, c, d) \mid a^2 + b^2 = 0\}$;
(d) $U = \{(a, b, c, d) \mid a^2 + b^2 = 1\}$;
(e) $U = \{(a + 2b, 0, 2a - b, b) \mid a, b \in \mathbb{R}\}$;
(f) $U = \{(a + 2b, a, 2a - b, b) \mid a, b \in \mathbb{R}\}$.

4.7 Which of the following criteria are correct for a non-empty subset U of a real vector space V to be a subspace of V?
(a) $(\forall x, y \in U)(\forall a, b \in \mathbb{R}) \quad ax + by \in U$;
(b) $(\forall x, y \in U)(\forall a \in \mathbb{R}) \quad ax + y \in U$;
(c) $(\forall x, y \in U)(\forall a \in \mathbb{R}) \quad ax + ay \in U$;
(d) $(\forall x, y \in U)(\forall a \in \mathbb{R}) \quad ax - ay \in U$.

4.8 Let V be a real vector space and suppose that $v_1, \ldots, v_k$ are linearly independent vectors in V. If $v = \sum_{i=1}^{k} a_i v_i$ where each $a_i \in \mathbb{R}$, prove that $v - v_1, \ldots, v - v_k$ are linearly independent if and only if $\sum_{i=1}^{k} a_i \neq 1$.

4.9 Let V be a vector space and let X, Y be subspaces of V. Suppose that $\dim V = 10$, $\dim X = 8$ and $\dim Y = 9$. What are the possible values of $\dim (X \cap Y)$?

4.10 Show that there are

$$2^{n(n-1)/2}(2^n - 1)(2^{n-1} - 1) \cdots (2^3 - 1)(2^2 - 1)$$

non-singular $n \times n$ matrices each of whose entries is 0 or 1.

4.11 Let $A_1, \ldots, A_k \in \mathrm{Mat}_{m\times n}(\mathbb{R})$. Let X and Y be invertible $m \times m$ and $n \times n$ matrices respectively, and let $B \in \mathrm{Mat}_{n\times p}(\mathbb{R})$. Prove that

(*a*) $\{A_1, \ldots, A_k\}$ is a linearly independent subset of $\mathrm{Mat}_{m\times n}(\mathbb{R})$ if and only if $\{XA_1Y, \ldots, XA_kY\}$ is a linearly independent subset of $\mathrm{Mat}_{m\times n}(\mathbb{R})$;

(*b*) if $\{A_1B, \ldots, A_kB\}$ is a linearly independent subset of $\mathrm{Mat}_{m\times p}(\mathbb{R})$ then $\{A_1, \ldots, A_k\}$ is a linearly independent subset of $\mathrm{Mat}_{m\times n}(\mathbb{R})$.

Give an example to show that $\{A_1, \ldots, A_k\}$ being linearly independent need not imply that $\{A_1B, \ldots, A_kB\}$ is linearly independent.

4.12 Find bases of $\mathrm{Mat}_{2\times 2}(\mathbb{Q})$ that consist

(*a*) of matrices A with $A^2 = A$;

(*b*) of invertible matrices;

(*c*) of matrices with determinant 1.

Show that it is impossible to find a basis of $\mathrm{Mat}_{2\times 2}(\mathbb{Q})$ consisting of commuting matrices. For which values of n is it possible to find such a basis of $\mathrm{Mat}_{n\times n}(\mathbb{Q})$?

4.13 Let V be a real vector space and let $b_1, b_2 \in V$ be linearly independent. If $a_1, a_2 \in V$ are given by

$$a_1 = \alpha_1 b_1 + \alpha_2 b_2, \quad a_2 = \beta_1 b_1 + \beta_2 b_2$$

where $\alpha_1, \alpha_2, \beta_1, \beta_2 \in \mathbb{R}$, prove that the subspace generated by $\{a_1, a_2\}$ coincides with that generated by $\{b_1, b_2\}$ if and only if

$$\det \begin{bmatrix} \alpha_1 & \alpha_2 \\ \beta_1 & \beta_2 \end{bmatrix} \neq 0.$$

4.14 Let U be the subspace of $\mathbb{R}^4$ spanned by

$$X = \{(2, 2, 1, 3), (7, 5, 5, 5), (3, 2, 2, 1), (2, 1, 2, 1)\}.$$

Given that $\boldsymbol{x} = (6 + \lambda, 1 + \lambda, -1 + \lambda, 2 + \lambda)$ belongs to U, find λ. For this value of λ, does $\boldsymbol{x}$ have a unique expression as a linear combination of the vectors of X?

Find a basis for U and extend this to a basis for $\mathbb{R}^4$.

4.15 Prove that $V = \{a + b\sqrt{2} \mid a, b \in \mathbb{Q}\}$ is a vector space over $\mathbb{Q}$.

4.16 Let V be the real vector space of continuous functions $f : \mathbb{R} \to \mathbb{R}$. Which of the following subsets are subspaces of V?

(*a*) $W_1 = \{f \in V \mid f(\frac{1}{2}) \in \mathbb{Q}\}$;

(*b*) $W_2 = \{f \in V \mid f(\frac{1}{2}) = f(1)\}$;
(*c*) $W_3 = \{f \in V \mid f(\frac{1}{2}) = 0\}$;
(*d*) $W_4 = \{f \in V \mid \mathrm{D}f(\frac{1}{2}) = 1\}$.

Find $W_i \cap W_j$ in the cases where W_i and W_j are subspaces.

4.17 Let $\mathbb{Z}_3$ be the field of integers modulo 3. Consider the $\mathbb{Z}_3$-vector space $\mathbb{Z}_3^3 = \{(a, b, c) \mid a, b, c \in \mathbb{Z}_3\}$. Which of the following subsets are linearly independent?

(*a*) $A_1 = \{(1, 2, 0), (2, 1, 0)\}$;
(*b*) $A_2 = \{(1, 1, 1), (1, 0, 1), (1, 0, 0), (0, 0, 1)\}$;
(*c*) $A_3 = \{(1, 2, 0), (1, 1, 1), (2, 0, 1)\}$;
(*d*) $A_4 = \{(1, 0, 1), (1, 1, 0), (0, 1, 1)\}$.

4.18 Prove that $\{(3 - \mathrm{i}, 2 + 2\mathrm{i}, 4), (2, 2 + 4\mathrm{i}, 3), (1 - \mathrm{i}, -2\mathrm{i}, -1)\}$ is a basis of the $\mathbb{C}$-vector space $\mathbb{C}^3$. Express each of $(1, 0, 0)$, $(0, 1, 0)$, $(0, 0, 1)$ as a linear combination of these basis vectors.

4.19 Consider the vector space $\mathbb{Z}_2^3$. How many elements does it have? How many different bases are there?

If V is a vector space of dimension 3 over $\mathbb{R}$ and if $\{x_1, x_2, x_3\}$ is a basis of V, prove that $\{x_1 + x_2, x_2 + x_3, x_3 + x_1\}$ and $\{x_1, x_1 + x_2, x_1 + x_2 + x_3\}$ are also bases of V. Is this still true if we replace V by $\mathbb{Z}_2^3$?

4.20 The set Map $(\mathbb{R}, \mathbb{R})$ of all mappings from $\mathbb{R}$ to $\mathbb{R}$ is a real vector space under addition and scalar multiplication. Let n be a positive integer and let E_n be the set of all mappings $f : \mathbb{R} \to \mathbb{R}$ given by a prescription of the form

$$f(x) = a_0 + \sum_{k=1}^{n} (a_k \cos kx + b_k \sin kx)$$

where a_i, $b_i \in \mathbb{R}$ for every i. Show that E_n is a subspace of the real vector space Map $(\mathbb{R}, \mathbb{R})$. If $f \in E_n$ is the zero map, prove that all the coefficients a_i, b_i must be 0.

(*Hint:* If D denotes the differentiation map, find a prescription for $\mathrm{D}^2 f + n^2 f$ and use induction.)

Deduce that the $2n + 1$ functions

$$x \to 1, \quad x \to \cos kx, \quad x \to \sin kx \quad (k = 1, \ldots, n)$$

form a basis for E_n.

4.21 Let $\alpha, \beta \in \mathbb{R}$ with $\alpha \neq \beta$ and let r, s be fixed positive integers. Show that the set of functions of the form

$$x \to f(x) = \frac{a_0 + a_1 x + \dots + a_{r+s-1}x^{r+s-1}}{(x-\alpha)^r (x-\beta)^s}$$

where each $a_i \in \mathrm{IR}$, is a real vector space of dimension $r + s$.

(*Hint:* Show that the functions

$$x \to f_i(x) = \frac{x^i}{(x-\alpha)^r (x-\beta)^s}$$

for $i = 0, \dots, r + s - 1$ constitute a basis.)

Show also that if g_i and h_j are given by $g_i(x) = (x - \alpha)^{-i}$ and $h_j(x) = (x - \beta)^{-j}$ then

$$B = \{g_i \mid i = 1, \dots, r\} \cup \{h_j \mid j = 1, \dots, s\}$$

is also a basis.

(*Hint:* It suffices to show that B is linearly independent.)

4.22 Let V be a finite-dimensional vector space over a field F. Prove that V has precisely one basis if and only if either (*a*) $V = \{0\}$, or (*b*) F has only two elements and $\dim V = 1$.

4.23 For each positive integer k let $f_k : \mathrm{IR} \to \mathrm{IR}$ be given by

$$f_k(x) = \exp r_k x \quad (r_k \in \mathrm{IR}).$$

Prove that $\{f_1, \dots, f_n\}$ is linearly independent if and only if $r_1, \dots, r_n$ are distinct.

5: Linear mappings

If V, W are vector spaces (over the same field) of dimensions m, n respectively and if $f: V \to W$ is a linear transformation (i.e. $f \in \text{Map}(V, W)$) then the matrix associated with f with respect to fixed ordered bases $\{v_1, \ldots, v_m\}$ of V and $\{w_1, \ldots, w_n\}$ of W is the matrix A whose ith column consists of the coordinates of $f(v_i)$ relative to the basis (w_i). If $f: V \to W$ is represented by the matrix B relative to ordered bases (v_i') and (w_i') then A and B are related by $B = Q^{-1}AP$ where P represents the identity map on V relative to the bases (v_i), (v_i') and Q represents the identity map on W relative to the bases (w_i), (w_i'). When A, B are $n \times n$ matrices and $(w_i) = (v_i)$, $(w_i') = (v_i')$ then $Q = P$ gives $B = P^{-1}AP$, in which case B is said to be similar to A.

Particular subspaces associated with a linear map $f: V \to W$ are the kernel $\text{Ker} f = \{x \in V \mid f(x) = 0\}$ and the image $\text{Im} f = \{f(x) \mid x \in V\}$. The dimension theorem states that

$$\dim V = \dim \text{Im} f + \dim \text{Ker} f.$$

The rank of f is defined to be $\dim \text{Im} f$. This coincides with the rank of any matrix that represents f. A linear map $f: V \to W$ is an isomorphism if it is both injective and surjective; equivalently, $\text{Im} f = W$ and $\text{Ker} f = \{0\}$.

5.1 Determine which of the following mappings $f: \mathbb{R}^3 \to \mathbb{R}^3$ are linear:

(*a*) $f(x, y, z) = (z, -y, x)$;
(*b*) $f(x, y, z) = (|x|, 0, -y)$;
(*c*) $f(x, y, z) = (y, z, 0)$;
(*d*) $f(x, y, z) = (x - 1, x, y)$;
(*e*) $f(x, y, z) = (2x, y - 2, 3y)$;
(*f*) $f(x, y, z) = (2x, y, 3y)$.

5.2 (*a*) If $f: \mathbb{R}^3 \to \mathbb{R}^3$ is linear and such that

$$f(1,1,1)=(1,1,1),\quad f(1,2,3)=(-1,-2,-3),$$
$$f(1,1,2)=(2,2,4)$$

is it possible to find $f(a,b,c)$ for all $(a,b,c)\in \mathbb{R}^3$?

(*b*) If $f:\mathbb{R}^3\to\mathbb{R}^3$ is linear and such that

$$f(1,1,1)=(1,1,1),\quad f(2,2,3)=(3,3,5),\quad f(1,1,2)=(2,2,4)$$

is it possible to find $f(a,b,c)$ for all $(a,b,c)\in \mathbb{R}^3$?

(*c*) Does there exist a linear mapping $f:\mathbb{R}^3\to\mathbb{R}^3$ with the property that

$$f(0,1,1)=(3,1,-2),\quad f(1,0,1)=(4,-1,1),$$
$$f(1,1,0)=(-3,2,1),\quad f(1,1,1)=(3,4,2)?$$

5.3 Let B be a fixed non-zero element of $\text{Mat}_{n\times n}(\mathbb{R})$. Which of the following mappings $T:\text{Mat}_{n\times n}(\mathbb{R})\to\text{Mat}_{n\times n}(\mathbb{R})$ are linear?

(*a*) $T(A)=AB-BA$;

(*b*) $T(A)=AB+BA$;

(*c*) $T(A)=AB^2+BA$;

(*d*) $T(A)=AB^2-BA^2$;

(*e*) $T(A)=(A+B)^2-(A+2B)(A-3B)$.

5.4 Find the matrices A and B associated with the linear mappings $f:\mathbb{R}^2\to\mathbb{R}^3$ and $g:\mathbb{R}^3\to\mathbb{R}^2$ with respect to the standard bases when f and g are given by

$$f(x,y)=(x+2y,2x-y,-x),\quad g(x,y,z)=(2x-y,2y-z).$$

Find the matrices C and D associated with these mappings f and g with respect to the bases $\{(0,1),(1,1)\}$ of $\mathbb{R}^2$ and $\{(0,0,1),(0,1,1),(1,1,1)\}$ of $\mathbb{R}^3$.

5.5 Suppose that the mapping $f:\mathbb{R}^3\to\mathbb{R}^3$ is linear and such that

$$f(1,0,0)=(2,3,-2),\quad f(1,1,0)=(4,1,4),\quad f(1,1,1)=(5,-1,7)$$

Find the matrix of f with respect to the standard basis of $\mathbb{R}^3$.

5.6 Find the matrix that represents the linear mapping $f:\mathbb{R}^3\to\mathbb{R}^3$ given by

$$f(a,b,c)=(2a+c,b-a+c,3c)$$

with respect to the basis $\{(1,-1,0),(1,0,-1),(1,0,0)\}$.

5.7 In $\mathbb{R}^2$ let M and N be distinct lines passing through the origin. Let $p_{M,N}:\mathbb{R}^2\to\mathbb{R}^2$ be the map such that, for every $(x,y)\in\mathbb{R}^2, p_{M,N}(x,y)$ is the point of intersection of M with the line through (x,y) parallel to N. Determine $p_{M,N}(x,y)$ in terms of

(*a*) the gradients m,n of M,N if neither is the y-axis;

(*b*) the gradient m of M if N is the y-axis;
(*c*) the gradient n of N if M is the y-axis.

Conclude that in all cases $p_{M,N}$ is linear. Determine the corresponding matrix relative to the standard basis of $\mathbb{R}^2$ in each of the above cases. What is the minimum polynomial of $p_{M,N}$?

5.8 If V, W are vector spaces over a field F prove that a linear transformation $f: V \to W$ is injective if and only if, whenever $\{v_1, \ldots, v_n\}$ is a linearly independent subset of V, $\{f(v_1), \ldots, f(v_n)\}$ is a linearly independent subset of W.

5.9 Given $x, y \in \mathbb{R}^3$ define the wedge product of x, y by

$$\begin{aligned} x \wedge y &= (x_1, x_2, x_3) \wedge (y_1, y_2, y_3) \\ &= (x_2 y_3 - x_3 y_2, x_3 y_1 - x_1 y_3, x_1 y_2 - x_2 y_1). \end{aligned}$$

Define $f_y : \mathbb{R}^3 \to \mathbb{R}^3$ by $f_y(x) = x \wedge y$. Show that f_y is linear and prove that if $y \neq 0$ then $\operatorname{Ker} f_y$ is the subspace of $\mathbb{R}^3$ generated by $\{y\}$.

(*Hint:* Without loss of generality, suppose that $y_1 \neq 0$. Consider separately the cases $y_2 = 0 = y_3, y_2 \neq 0 = y_3, y_2 = 0 \neq y_3, y_2 \neq 0 \neq y_3$.)

5.10 If $f : \mathbb{R} \to \mathbb{R}$ is linear and $\vartheta : \mathbb{R}^2 \to \mathbb{R}^2$ is given by

$$\vartheta(x, y) = (x, y - f(x)),$$

prove that ϑ is an isomorphism.

5.11 Let V be a vector space of dimension n over a field F. If $f : V \to V$ is linear, prove that the following are equivalent:

(*a*) $\operatorname{Im} f = \operatorname{Ker} f$;
(*b*) $f \circ f = 0, f \neq 0$, n is even and $\operatorname{rank} f = \frac{1}{2}n$.

5.12 A diagram of finite-dimensional vector spaces and linear transformations of the form

$$V_1 \xrightarrow{f_1} V_2 \xrightarrow{f_2} V_3 \xrightarrow{f_3} \cdots \xrightarrow{f_n} V_{n+1}$$

is called an exact sequence if

(*a*) f_1 is injective;
(*b*) f_n is surjective;
(*c*) $(i = 1, \ldots, n-1)$ $\operatorname{Im} f_i = \operatorname{Ker} f_{i+1}$.

Prove that, for such an exact sequence,

$$\sum_{i=1}^{n+1} (-1)^i \dim V_i = 0.$$

5.13 Let V be a vector space of dimension n over a field F. A linear transformation $f : V \to V$ is said to be nilpotent if $f^p = 0$ for some positive integer p. The smallest such integer p is called the index of nilpotency of f.

Suppose that f is nilpotent of index p. If $x \in V$ is such that $f^{p-1}(x) \neq 0$ prove that

$$\{x, f(x), f^2(x), \ldots, f^{p-1}(x)\}$$

is linearly independent. Hence show that f is nilpotent of index n if and only if there is an ordered basis $(a_i)_n$ of V such that the matrix of f relative to $(a_i)_n$ is of the form

$$\begin{bmatrix} 0 & 0 & 0 & \ldots & 0 & 0 \\ 1 & 0 & 0 & \ldots & 0 & 0 \\ 0 & 1 & 0 & \ldots & 0 & 0 \\ 0 & 0 & 1 & \ldots & 0 & 0 \\ \vdots & \vdots & \vdots & & \vdots & \vdots \\ 0 & 0 & 0 & \ldots & 1 & 0 \end{bmatrix}.$$

Deduce that an $n \times n$ matrix M over F is such that $M^n = 0$ and $M^{n-1} \neq 0$ if and only if M is similar to a matrix of the above form.

5.14 Given $\vartheta \in \mathbb{R}$ prove that the following matrices over $\mathbb{C}$ are similar:

$$\begin{bmatrix} \cos\vartheta & -\sin\vartheta \\ \sin\vartheta & \cos\vartheta \end{bmatrix}, \quad \begin{bmatrix} e^{i\vartheta} & 0 \\ 0 & e^{-i\vartheta} \end{bmatrix}.$$

5.15 Which of the following mappings $f : \mathbb{Z}_2^3 \to \mathbb{Z}_2^3$ are linear?

(*a*) $f(a,b,c) = (b, -a, 1)$;
(*b*) $f(a,b,c) = (a^2, b^2, c^2)$;
(*c*) $f(a,b,c) = (a+1, b+1, c+1)$.

5.16 Let $f : \mathbb{Z}_2^2 \to \mathbb{Z}_2^2$ be such that

$$f(0,0) = (0,0), \quad f(1,0) = (1,1), \quad f(0,1) = (1,0), \quad f(1,1) = (0,1)$$

Is f linear?

5.17 Let $\mathbb{R}[X]$ denote the vector space of polynomials over $\mathbb{R}$. Prove that the differentiation map $\mathrm{D} : \mathbb{R}[X] \to \mathbb{R}[X]$ is linear. Prove also that the map $T : \mathbb{R}[X] \to \mathbb{R}[X]$ given by $T(p(X)) = Xp(X)$ is linear. Describe the linear maps $\mathrm{D} + T$, $\mathrm{D} \circ T$ and $T \circ \mathrm{D}$. Show that $\mathrm{D} \circ T - T \circ \mathrm{D} = I$ where I is the identity map. Deduce that $(T \circ \mathrm{D})^2 = T^2 \circ \mathrm{D}^2 + T \circ \mathrm{D}$. Find a linear map

$S : \mathbb{R}[X] \to \mathbb{R}[X]$ such that $D \circ S = I$. Show that neither D nor S is an isomorphism.

5.18 Let F be a field and define $f : F^3 \to F^3$ by

$$f(a, b, c) = (a + b, a + c, b + c).$$

Find $\operatorname{Im} f$ and $\operatorname{Ker} f$ when

(*a*) $F = \mathbb{R}$;

(*b*) $F = \mathbb{Z}_2$.

What are the dimensions of $\operatorname{Im} f$ and $\operatorname{Ker} f$ in each of these cases?

5.19 Consider the linear transformation $f : \mathbb{R}^3 \to \mathbb{R}^3$ given by

$$f(a, b, c) = (b, -a, c).$$

Find the matrix A of f with respect to the basis

$$\{(1, 0, 0), (0, 1, 0), (0, 0, 1)\}$$

and the matrix B of f with respect to the basis

$$\{(1, 1, 0), (0, 1, 1), (1, 0, 1)\}.$$

Determine a non-singular matrix X such that $A = X^{-1}BX$.

5.20 Let V be a vector space over $\mathbb{R}$ with basis $\{u_1, u_2\}$. Let $f : V \to V$ be the linear transformation such that

$$f(u_1) = u_1 - u_2, \quad f(u_2) = u_1.$$

Find the matrix of f with respect to the basis $\{u_1, u_2\}$. Find also the matrix of f with respect to the basis $\{w_1, w_2\}$ where

$$w_1 = 3u_1 - u_2, \quad w_2 = u_1 + u_2.$$

5.21 For $n \geqslant 1$ let $\mathbb{R}_n[X]$ be the vector space of real polynomials of degree less than or equal to $n - 1$. Determine the matrix of the differentiation map D with respect to the bases

(*a*) $\{1, X, X^2, \ldots, X^{n-1}\}$;

(*b*) $\{X^{n-1}, X^{n-2}, \ldots, X, 1\}$;

(*c*) $\{1, 1 + X, 1 + X^2, \ldots, 1 + X^{n-1}\}$.

5.22 Prove that $\{1, i\}$ is a basis for the real vector space $\mathbb{C}$. Show that $f : \mathbb{C} \to \mathbb{C}$ given by

$$f(x + yi) = x - yi$$

is linear and find the matrix of f relative to the basis $\{1, i\}$.

Show that f is not linear when $\mathbb{C}$ is regarded as a vector space over $\mathbb{C}$.

5.23 Let V be a vector space of dimension 3 over a field F and let $f : V \to V$ be a linear transformation. Suppose that, relative to some basis of V, the matrix of f is

$$\begin{bmatrix} 3 & -1 & 1 \\ -1 & 5 & -1 \\ 1 & -1 & 3 \end{bmatrix}.$$

Determine $\dim \operatorname{Im} f$ and $\dim \operatorname{Ker} f$ when

(*a*) $F = \mathbb{R}$;

(*b*) $F = \mathbb{Z}_2$;

(*c*) $F = \mathbb{Z}_3$.

5.24 Give an example of a vector space V and a non-zero non-identity linear transformation $f : V \to V$ such that

$$\operatorname{Im} f \cap \operatorname{Ker} f \neq \{0\}.$$

Is it possible to have $\operatorname{Im} f = \operatorname{Ker} f$? What about $\operatorname{Im} f \subset \operatorname{Ker} f$? What about $\operatorname{Ker} f \subset \operatorname{Im} f$?

5.25 Let p be a prime and let V be a vector space of dimension 3 over the field $\mathbb{Z}_p$. Suppose that $f : V \to V$ is a linear transformation such that, relative to some basis of V, the matrix of f is

$$\begin{bmatrix} 2 & 2 & 1 \\ 1 & 3 & 1 \\ 1 & 2 & 2 \end{bmatrix}.$$

Prove that f is an isomorphism if and only if $p \neq 5$.

6: Inner product spaces

Standard notation will be used for inner products, namely $\langle x \mid y \rangle$. If $x, y \in \mathrm{IR}^n$ the standard inner product is given by $\langle x \mid y \rangle = \sum_{i=1}^{n} x_i y_i$ where x_i, y_i are the ith components of x, y. The length of a vector x is $\|x\| = \sqrt{\langle x \mid x \rangle}$. Vectors x, y are orthogonal if $\langle x \mid y \rangle = 0$. A set $\{x_1, \ldots, x_n\}$ is orthonormal if $\langle x_i \mid x_j \rangle = \delta_{ij}$. The Gram-Schmidt orthonormalisation process produces an orthonormal basis for the subspace spanned by a given linearly independent set.

For a subspace W of an inner product space V the orthogonal complement of W is the subspace

$$W^\perp = \{y \in V \mid (\forall x \in W)\, \langle x \mid y \rangle = 0\}.$$

For a given linear map $f : V \to V$ it can be shown that there is a unique linear map $f^* : V \to V$ such that

$$\langle f(x) \mid y \rangle = \langle x \mid f^*(y) \rangle.$$

This is called the adjoint of f. We say that f is self-adjoint if $f = f^*$, and normal if $f \circ f^* = f^* \circ f$.

6.1 If V is a real inner product space and $x, y \in V$ prove that

$$\left\langle y - \frac{\langle x \mid y \rangle}{\|x\|^2} x \,\middle|\, x \right\rangle = 0.$$

Deduce that if $x, y \in \mathrm{IR}^3$ with $\{x, y\}$ linearly independent, and if L is the line through the origin and x, then the line through y perpendicular to L meets L at the point $(\langle x \mid y \rangle / \|x\|^2)x$. If ϑ is the angle xOy prove that $\cos \vartheta = \langle x \mid y \rangle / \|x\| \, \|y\|$.

6.2 If V is a real inner product space and $x, y \in V$ prove that

$$\|x + y\|^2 = \|x\|^2 + \|y\|^2 + 2\langle x \mid y \rangle.$$

Interpret this result in $\mathbb{R}^2$. If W is a complex inner product space and $x, y \in W$ prove that

$$\|x+y\|^2 - \mathrm{i}\|\mathrm{i}x+y\|^2 = \|x\|^2 + \|y\|^2 - \mathrm{i}(\|x\|^2 + \|y\|^2) + 2\langle x \mid y\rangle.$$

6.3 If V is an inner product space, establish the parallelogram identity

$$(\forall x, y \in V) \quad \|x+y\|^2 + \|x-y\|^2 = 2\|x\|^2 + 2\|y\|^2.$$

Interpret this result in $\mathbb{R}^2$.

6.4 If V is a real inner product space and $x, y \in V$ are such that $\|x\| = \|y\|$ show that $\langle x+y \mid x-y\rangle = 0$. Interpret this result in $\mathbb{R}^2$.

6.5 Write down the Cauchy-Schwarz inequality ($|\langle u \mid v\rangle| \leqslant \|u\| \,.\, \|v\|$) when the inner product space in question is

(*a*) $\mathbb{C}^n$ under the standard inner product;

(*b*) the vector space of continuous functions $f: [0,1] \to \mathbb{R}$ under the inner product defined by

$$\langle f \mid g\rangle = \int_0^1 f(t)\,g(t)\,\mathrm{d}t.$$

6.6 Let V denote the vector space $\mathrm{Mat}_{n\times n}(\mathbb{C})$. For every $A = [a_{ij}] \in V$ define the conjugate transpose of A to be the matrix A^* the (i,j)th element of which is $\bar{a}_{ji}$; and define the trace of A by $\mathrm{tr}\,(A) = \Sigma_{i=1}^n a_{ii}$.

Show that $\mathrm{tr}\,(A+B) = \mathrm{tr}\,(A) + \mathrm{tr}\,(B)$ and $\mathrm{tr}\,(AB) = \mathrm{tr}\,(BA)$ for all $A, B \in V$, and that $\langle A \mid B\rangle = \mathrm{tr}\,(AB^*)$ defines an inner product on V. Interpret the Cauchy-Schwarz inequality in this complex inner product space.

Show further that if $E_{pq} \in V$ is the matrix whose (p,q)th element is 1 and all other elements are 0 then

$$\{E_{pq} \mid p, q = 1, \ldots, n\}$$

is an orthonormal basis of V.

6.7 Use the Gram-Schmidt orthonormalisation process to find an orthonormal basis for the subspace of $\mathbb{R}^4$ generated by

$$\{(1,1,0,1), (1,-2,0,0), (1,0,-1,2)\}.$$

6.8 Consider the real inner product space of continuous functions $f: [0,1] \to \mathbb{R}$ under the inner product

$$\langle f \mid g\rangle = \int_0^1 f(t)\,g(t)\,\mathrm{d}t.$$

Find an orthonormal basis for the subspace generated by

$$\{t \to 1, t \to t\}.$$

6.9 Let V and W be subspaces of a finite-dimensional inner product space. Prove that, if $^{\perp}$ denotes orthogonal complement,

(a) $(V \cap W)^{\perp} = V^{\perp} + W^{\perp}$;

(b) $(V + W)^{\perp} = V^{\perp} \cap W^{\perp}$.

6.10 Let $\mathrm{IR}_3[X]$ be the real vector space of polynomials of degree less than or equal to 2. Show that the prescription

$$\langle p \mid q \rangle = \int_0^1 p(X)\, q(X)\, \mathrm{d}X$$

defines an inner product on $\mathrm{IR}_3[X]$. If K denotes the subspace of constant polynomials find the orthogonal complement $K^{\perp}$ of K in $\mathrm{IR}_3[X]$. Write down a basis for $K^{\perp}$ and use the Gram–Schmidt orthonormalisation process to compute an orthonormal basis for $K^{\perp}$.

6.11 Let V be the complex inner product space $\mathrm{Mat}_{n \times n}(\mathbb{C})$ (see Exercise 6.6). For every $M \in V$ let $f_M : V \to V$ be given by $f_M(A) = MA$. Prove that the adjoint of f_M is given by $f_M^* = f_{M^*}$.

6.12 Let V be the real inner product space of polynomials over IR with inner product $\langle p \mid q \rangle = \int_0^1 p(X) q(X)\, \mathrm{d}X$. For every $p \in V$ let $f_p : V \to V$ be given by $(\forall q \in V)\quad f_p(q) = pq$ where $(pq)(X) = p(X)\, q(X)$. Prove that f_p is self-adjoint.

6.13 If V is a finite-dimensional inner product space and $f : V \to V$ is linear prove that

$$\operatorname{Im} f^* = (\operatorname{Ker} f)^{\perp} \quad \text{and} \quad \operatorname{Ker} f^* = (\operatorname{Im} f)^{\perp}.$$

6.14 Let V be a complex inner product space. For all $x, y \in V$ let $f_{x,y} : V \to V$ be given by

$$f_{x,y}(z) = \langle z \mid y \rangle x.$$

Prove that $f_{x,y}$ is linear and that

(a) $(\forall x, y, z \in V)\quad f_{x,y} \circ f_{y,z} = \|y\|^2 f_{x,z}$;

(b) the adjoint of $f_{x,y}$ is $f_{y,x}$.

Hence show that if $x \neq 0$ and $y \neq 0$ then

(c) $f_{x,y}$ is normal $\Leftrightarrow (\exists \lambda \in \mathbb{C})\quad x = \lambda y$;

(d) $f_{x,y}$ is self-adjoint $\Leftrightarrow (\exists \lambda \in \mathrm{IR})\quad x = \lambda y$.

Solutions to Chapter 1

1.1 The respective products are

$$\begin{bmatrix} 1 & 7 \\ 3 & 1 \end{bmatrix}; \quad \begin{bmatrix} 3 & 7 \\ 2 & 5 \\ 2 & 3 \end{bmatrix}; \quad \begin{bmatrix} 4 & 0 & 0 \\ 0 & 4 & 0 \\ 0 & 0 & 4 \end{bmatrix};$$

$$\begin{bmatrix} 1 & 2 & 3 & 4 \\ 2 & 4 & 6 & 8 \\ 3 & 6 & 9 & 12 \\ 4 & 8 & 12 & 16 \end{bmatrix}; \quad [30].$$

1.2 The product is the 1×1 matrix $[t]$ where

$$\begin{aligned} t &= x(ax + hy + g) + y(hx + by + f) + gx + fy + c \\ &= ax^2 + 2hxy + by^2 + 2gx + 2fy + c. \end{aligned}$$

Denoting by 0 the 1×1 zero matrix, we can write each of the equations the form

$$[x \quad y \quad 1]M\begin{bmatrix} x \\ y \\ 1 \end{bmatrix} = 0$$

where

$$(a)\ M = \begin{bmatrix} 1 & \frac{9}{2} & 4 \\ \frac{9}{2} & 1 & \frac{5}{2} \\ 4 & \frac{5}{2} & 2 \end{bmatrix}; \quad (b)\ M = \begin{bmatrix} \dfrac{1}{\alpha^2} & 0 & 0 \\ 0 & \dfrac{1}{\beta^2} & 0 \\ 0 & 0 & -1 \end{bmatrix};$$

$$(c)\ M=\begin{bmatrix}0&\frac{1}{2}&0\\ \frac{1}{2}&0&0\\ 0&0&-\alpha^2\end{bmatrix};\quad (d)\ M=\begin{bmatrix}0&0&-2\alpha\\ 0&1&0\\ -2\alpha&0&0\end{bmatrix}.$$

1.3 We have that $AB=0, A^2=A, B^2=-B$ and $(A+B)^2=I_2$. But

$$A^2+2AB+B^2=\begin{bmatrix}1&2\\0&1\end{bmatrix}.$$

Since $(A+B)^2=I_2$ we have $(A+B)^3=(A+B)^2(A+B)=A+B$; and since $AB=0, A^3=A, B^3=B$ we have

$$A^3+3A^2B+3AB^2+B^3=A^3+B^3=A+B.$$

1.4 A simple calculation shows that

$$A^2=\begin{bmatrix}0&0&a^2&2a^3\\0&0&0&a^2\\0&0&0&0\\0&0&0&0\end{bmatrix},\quad A^3=\begin{bmatrix}0&0&0&a^3\\0&0&0&0\\0&0&0&0\\0&0&0&0\end{bmatrix}$$

and $A^4=0$, whence $A^n=0$ for every $n\geqslant 4$. It follows that

$$B=A-\tfrac{1}{2}A^2+\tfrac{1}{3}A^3=\begin{bmatrix}0&a&\frac{1}{2}a^2&\frac{1}{3}a^3\\0&0&a&\frac{1}{2}a^2\\0&0&0&a\\0&0&0&0\end{bmatrix}.$$

A simple calculation shows that

$$B^2=\begin{bmatrix}0&0&a^2&a^3\\0&0&0&a^2\\0&0&0&0\\0&0&0&0\end{bmatrix},\quad B^3=\begin{bmatrix}0&0&0&a^3\\0&0&0&0\\0&0&0&0\\0&0&0&0\end{bmatrix}$$

and $B^4=0$, whence $B^n=0$ for every $n\geqslant 4$. Then

$$B+\frac{1}{2!}B^2+\frac{1}{3!}B^3=\begin{bmatrix}0&a&a^2&a^3\\0&0&a&a^2\\0&0&0&a\\0&0&0&0\end{bmatrix}=A.$$

1.5 Let $X = \begin{bmatrix} a & b \\ c & d \end{bmatrix}$. Then

$$X^2 = \begin{bmatrix} a^2 + bc & b(a+d) \\ c(a+d) & cb + d^2 \end{bmatrix}.$$

Thus $X^2 = I_2$ if and only if

$$a^2 + bc = 1,$$
$$b(a+d) = 0,$$
$$c(a+d) = 0,$$
$$cb + d^2 = 1.$$

Suppose that $b = 0$: then these equations reduce to

$$a^2 = 1,$$
$$c(a+d) = 0,$$
$$d^2 = 1,$$

from which we see that

either $a = d = 1, c = 0$;
or $a = d = -1, c = 0$;
or $a = 1, d = -1, c$ arbitrary;
or $a = -1, d = 1, c$ arbitrary.

Suppose now that $b \neq 0$: then we must have $a + d = 0$ whence we see that $d = -a$ and $c = (1 - a^2)/b$.

Thus we see that the possibilities for X are

$$\begin{bmatrix} 1 & 0 \\ 0 & 1 \end{bmatrix}, \quad \begin{bmatrix} -1 & 0 \\ 0 & -1 \end{bmatrix},$$

$$\begin{bmatrix} 1 & 0 \\ c & -1 \end{bmatrix}, \quad \begin{bmatrix} -1 & 0 \\ c & 1 \end{bmatrix} \quad (c \in \mathbb{R}),$$

$$\begin{bmatrix} a & b \\ (1-a^2)/b & -a \end{bmatrix} \quad (a \in \mathbb{R}, b \in \mathbb{R} \setminus \{0\}).$$

1.6 Let $A = \begin{bmatrix} a & 0 \\ 0 & b \end{bmatrix}$. Then $A^2 = -I_2$ if and only if

$$a^2 + 1 = b^2 + 1 = 0.$$

Hence the complex diagonal solutions are

$$\begin{bmatrix} \mathrm{i} & 0 \\ 0 & -\mathrm{i} \end{bmatrix}, \begin{bmatrix} -\mathrm{i} & 0 \\ 0 & \mathrm{i} \end{bmatrix}, \begin{bmatrix} \mathrm{i} & 0 \\ 0 & \mathrm{i} \end{bmatrix}, \begin{bmatrix} -\mathrm{i} & 0 \\ 0 & -\mathrm{i} \end{bmatrix}.$$

No real diagonal solutions exist since $x^2 + 1 = 0$ has no real solutions.

If (x, y) lies on the curve $y^2 - x^2 = 1$ then we have

$$\begin{bmatrix} x & y \\ -y & -x \end{bmatrix} \begin{bmatrix} x & y \\ -y & -x \end{bmatrix} = -I_2$$

so there are infinitely many real 2×2 solutions.

If now A is a real 2×2 matrix such that $A^2 = -I_2$ then the $2k \times 2k$ matrix

$$\begin{bmatrix} A & & & \\ & A & & \\ & & \ddots & \\ & & & A \end{bmatrix}$$

is a solution of $X^2 = -I_{2k}$. Thus there are infinitely many $2k \times 2k$ real solutions of $X^2 = -I_{2k}$. As there is no real 1×1 solution, the result does not hold for all odd integers.

1.7 If $X = \begin{bmatrix} x & y \\ z & t \end{bmatrix}$ is such that $X^2 = 0$ then we have

$$\begin{bmatrix} x^2 + yz & xy + yt \\ xz + zt & yz + t^2 \end{bmatrix} = \begin{bmatrix} 0 & 0 \\ 0 & 0 \end{bmatrix}.$$

Consider therefore the equations

$$\begin{aligned} x^2 + yz &= 0, \\ y(x + t) &= 0, \\ z(x + t) &= 0, \\ yz + t^2 &= 0. \end{aligned}$$

Suppose that $y = 0$. Then clearly $x = t = 0$ and

$$X = \begin{bmatrix} 0 & 0 \\ z & 0 \end{bmatrix}$$

which is of the required form. Suppose now that $y \neq 0$. Then $t = -x$ and $x^2 + yz = 0$ which gives

$$X = \begin{bmatrix} x & y \\ -x^2/y & -x \end{bmatrix}.$$

Writing $a = \sqrt{y}$ and $b = x/\sqrt{y}$ we see that X is again of the required form.

The result fails over IR since, for example,

$$\begin{bmatrix} 0 & 0 \\ 1 & 0 \end{bmatrix}\begin{bmatrix} 0 & 0 \\ 1 & 0 \end{bmatrix} = \begin{bmatrix} 0 & 0 \\ 0 & 0 \end{bmatrix}$$

and there is no $b \in \mathbb{R}$ such that $-b^2 = 1$.

1.8 The condition is necessary since

$$\begin{bmatrix} x \\ y \end{bmatrix} [u \quad v] = \begin{bmatrix} xu & xv \\ yu & yv \end{bmatrix}$$

and clearly $xu \cdot yv = yu \cdot xv$.

Conversely, suppose that $X = \begin{bmatrix} a & b \\ c & d \end{bmatrix}$ is such that $ad = bc$. If $a \neq 0$ then define $x = a, y = c, u = 1, v = b/a$ and we have the required decomposition. If now $a = 0$ then either $b = 0$ or $c = 0$. If $b = 0$ we have

$$\begin{bmatrix} 0 & 0 \\ c & d \end{bmatrix} = \begin{bmatrix} 0 \\ 1 \end{bmatrix} [c \quad d]$$

and if $c = 0$ we have

$$\begin{bmatrix} 0 & b \\ 0 & d \end{bmatrix} = \begin{bmatrix} b \\ d \end{bmatrix} [0 \quad 1].$$

1.9 (*a*) (i) We have

$$\begin{aligned} [[AB]C] &= (AB - BA)C - C(AB - BA) \\ &= ABC + CBA - BAC - CAB; \\ [[BC]A] &= (BC - CB)A - A(BC - CB) \\ &= BCA + ACB - CBA - ABC; \\ [[CA]B] &= (CA - AC)B - B(CA - AC) \\ &= CAB + BAC - ACB - BCA. \end{aligned}$$

The result follows by adding these together.

(ii) $[(A + B)C] = (A + B)C - C(A + B) = AC - CA + BC - CB = [AC] + [BC]$.

(iii) Expand as in (i).

(*b*) Take, for example,

$$A = B = \begin{bmatrix} 0 & 1 \\ 1 & 0 \end{bmatrix}, \quad C = \begin{bmatrix} 1 & 0 \\ 0 & 0 \end{bmatrix}.$$

We have that $[[AB]C] = 0$ and

$$[A\,[BC]] = \begin{bmatrix} 2 & 0 \\ 0 & -2 \end{bmatrix}.$$

1.10 (*a*) Involves straightforward matrix multiplication.

(*b*) It is readily seen that

$$XY - YX = Z, \quad YZ - ZY = X, \quad ZX - XZ = Y.$$

(*c*) Involves straightforward multiplication.

1.11 (*a*) If $A = [a_{ij}]_{2\times 2}$ and $B = [b_{ij}]_{2\times 2}$ then it is readily seen that

$$AB - BA = \begin{bmatrix} a_{12}b_{21} - a_{21}b_{12} & ? \\ ? & a_{21}b_{12} - a_{12}b_{21} \end{bmatrix}$$

so the sum of the diagonal elements is 0. (There is no point in calculating the off-diagonal terms.)

(*b*) E is necessarily of the form

$$E = \begin{bmatrix} a_{11} & a_{12} \\ a_{21} & -a_{11} \end{bmatrix}$$

whence we see that

$$E^2 = \begin{bmatrix} a_{11}^2 + a_{12}a_{21} & a_{11}a_{12} - a_{11}a_{12} \\ a_{11}a_{21} - a_{11}a_{21} & a_{12}a_{21} + a_{11}^2 \end{bmatrix} = (a_{11}^2 + a_{12}a_{21})I_2.$$

(*c*) Let $E = AB - BA$. Then by (*a*) the sum of the diagonal elements of E is zero; and by (*b*) there is a scalar λ such that $E^2 = \lambda I_2$. Consequently

$$(AB - BA)^2 C = (\lambda I_2)C = \lambda C = C(\lambda I_2) = C(AB - BA)^2.$$

1.12 $A^t = A$ and $B^t = -B$. Thus

$$\begin{aligned}(AB + BA)^t &= (AB)^t + (BA)^t = B^tA^t + A^tB^t \\ &= -BA - AB = -(AB + BA)\end{aligned}$$

so that $AB + BA$ is skew-symmetric; similarly $AB - BA$ is symmetric. Next, $(A^2)^t = (A^t)^2 = A^2$ so A^2 is symmetric; similarly B^2 is symmetric. Finally,

$$(A^pB^qA^p)^t = (A^t)^p(B^t)^q(A^t)^p = A^p(-B)^qA^p$$

so $A^pB^qA^p$ is symmetric if q is even, and is skew-symmetric if q is odd.

1.13 $A^t = (xy^t - yx^t)^t = yx^t - xy^t = -A$ so A is skew-symmetric. If $x = [x_1 x_2 \cdots x_n]^t$ and $y = [y_1 y_2 \cdots y_n]^t$ then $x^ty = \sum_{i=1}^n x_iy_i = y^tx$. If now $x^tx = y^ty = [1]$ and $x^ty = y^tx = [k]$ then

$$A^2 = (xy^t - yx^t)(xy^t - yx^t)$$
$$= xy^txy^t - yx^txy^t - xy^tyx^t + yx^tyx^t$$
$$= kxy^t - yy^t - xx^t + kyx^t$$

and hence

$$A^3 = (kxy^t - yy^t - xx^t + kyx^t)(xy^t - yx^t)$$
$$= kxy^txy^t - yy^txy^t - xx^txy^t + kyx^txy^t$$
$$- kxy^tyx^t + yy^tyx^t + xx^tyx^t - kyx^tyx^t$$
$$= k^2xy^t - kyy^t - xy^t + kyy^t - kxx^t + yx^t + kxx^t - k^2yx^t$$
$$= k^2(xy^t - yx^t) - xy^t + yx^t$$
$$= (k^2 - 1)A.$$

1.14 Since $(AB)^t = B^tA^t$ we have $(A^t)^2 = (A^2)^t = A^t$. Also,

$$(A - A^t)^2 = A^2 - AA^t - A^tA + (A^t)^2,$$

so $(A - A^t)^2 = 0$ implies $A - AA^t - A^tA + A^t = 0$. Multiplying on the left by AA^t we obtain

$$AA^tA - (AA^t)^2 - AA^tA^tA + AA^tA^t = 0$$

which, since $A^tA^t = A^t$, reduces to $(AA^t)^2 = AA^t$.

1.15 Referring to Fig. S1.1, we have that

$$x' = r\cos(\alpha - \vartheta) = r\cos\alpha\cos\vartheta + r\sin\alpha\sin\vartheta = x\cos\vartheta + y\sin\vartheta$$
$$y' = r\sin(\alpha - \vartheta) = r\sin\alpha\cos\vartheta - r\cos\alpha\sin\vartheta = y\cos\vartheta - x\sin\vartheta$$

In matrix form, these equations become

$$\begin{bmatrix} x' \\ y' \end{bmatrix} = \begin{bmatrix} \cos\vartheta & \sin\vartheta \\ -\sin\vartheta & \cos\vartheta \end{bmatrix} \begin{bmatrix} x \\ y \end{bmatrix}.$$

Fig.S1.1

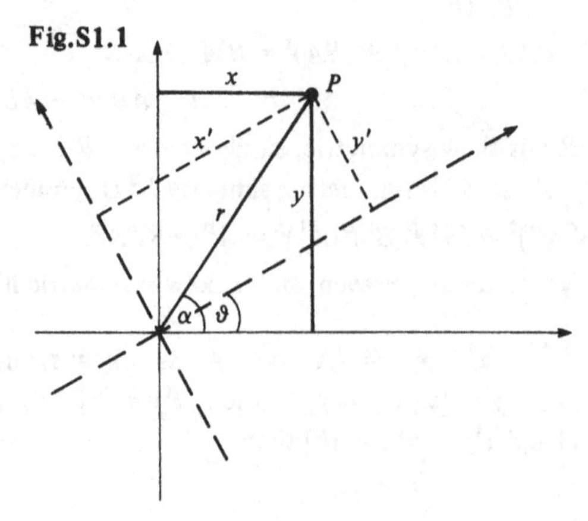

A rotation through ϑ followed by a rotation through φ is clearly equivalent to a rotation through $\vartheta + \varphi$. Consequently we have that

$$R_\vartheta R_\varphi = R_{\vartheta+\varphi} = R_\varphi R_\vartheta.$$

Rotating the hyperbola anti-clockwise through 45° is equivalent to rotating the axes clockwise through 45°. Thus we have

$$\begin{bmatrix} x' \\ y' \end{bmatrix} = R_{-\pi/4} \begin{bmatrix} x \\ y \end{bmatrix}.$$

Now since $R_{\pi/4}R_{-\pi/4} = R_{\pi/4-\pi/4} = R_0 = I_2$ we can multiply the above equation by $R_{\pi/4}$ to obtain

$$\begin{bmatrix} x \\ y \end{bmatrix} = R_{\pi/4} \begin{bmatrix} x' \\ y' \end{bmatrix} = \begin{bmatrix} \frac{1}{\sqrt{2}} & \frac{1}{\sqrt{2}} \\ -\frac{1}{\sqrt{2}} & \frac{1}{\sqrt{2}} \end{bmatrix} \begin{bmatrix} x' \\ y' \end{bmatrix}$$

so we have

$$x = \frac{1}{\sqrt{2}}x' + \frac{1}{\sqrt{2}}y'$$

$$y = -\frac{1}{\sqrt{2}}x' + \frac{1}{\sqrt{2}}y'$$

whence $x^2 - y^2 = 1$ becomes

$$\left(\frac{1}{\sqrt{2}}x' + \frac{1}{\sqrt{2}}y'\right)^2 + \left(-\frac{1}{\sqrt{2}}x' + \frac{1}{\sqrt{2}}y'\right)^2 = 1$$

i.e. $2x'y' = 1$. Thus the new equation of the hyperbola is $xy = \frac{1}{2}$. (See Fig. S1.2.)

Fig.S1.2

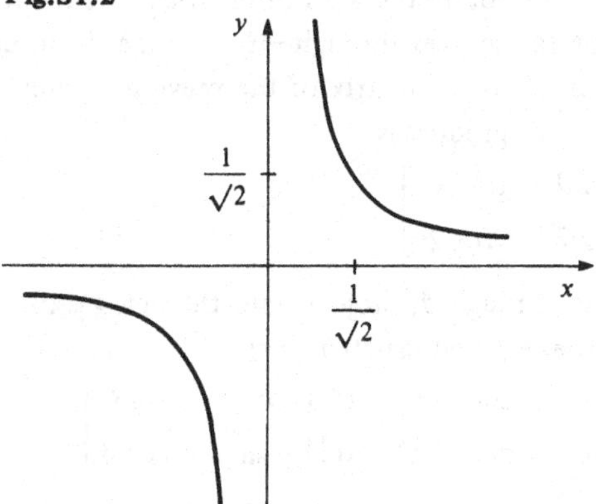

1.16 Let the top sheet be the (x,y)-plane and the bottom sheet be the (x',y')-plane. We have, where ϑ is the anti-clockwise angle of rotation of the top sheet,

$$\begin{bmatrix} \frac{5}{13} \\ \frac{12}{13} \end{bmatrix} = R_{-\vartheta} \begin{bmatrix} 1 \\ 0 \end{bmatrix} = \begin{bmatrix} \cos\vartheta & -\sin\vartheta \\ \sin\vartheta & \cos\vartheta \end{bmatrix} \begin{bmatrix} 1 \\ 0 \end{bmatrix} = \begin{bmatrix} \cos\vartheta \\ \sin\vartheta \end{bmatrix}$$

so $\cos\vartheta = \frac{5}{13}$ and $\sin\vartheta = \frac{12}{13}$ (see Fig. S1.3).

Fig.S1.3

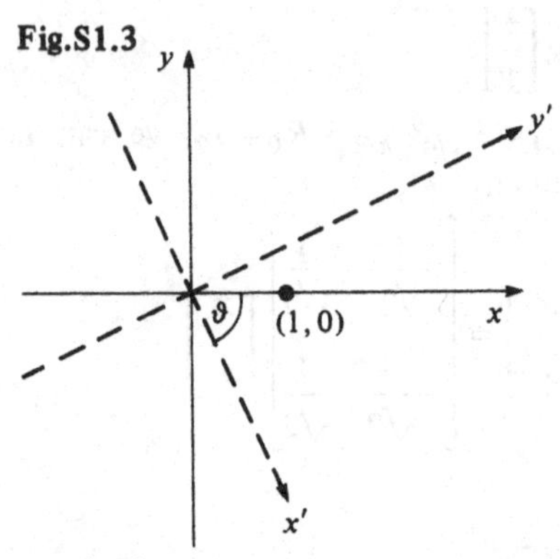

The point (x',y') above which the point $(2, 3)$ lies is then given by

$$\begin{bmatrix} x' \\ y' \end{bmatrix} = \begin{bmatrix} \frac{5}{13} & -\frac{12}{13} \\ \frac{12}{13} & \frac{5}{13} \end{bmatrix} \begin{bmatrix} 2 \\ 3 \end{bmatrix}$$

i.e. $(x',y') = (-2, 3)$.

1.17 Clearly $x' = x$ and $y' = -y$, so

$$\begin{bmatrix} x' \\ y' \end{bmatrix} = \begin{bmatrix} 1 & 0 \\ 0 & -1 \end{bmatrix} \begin{bmatrix} x \\ y \end{bmatrix}.$$

1.18 To obtain (x_L, y_L), rotate the axes through ϑ, take the reflection in the ne[w] x-axis, then rotate the axes through $-\vartheta$. The matrix in question is therefo[re] $R_{-\vartheta}MR_{\vartheta}$ where M is the matrix of the previous exercise. A simple calcu[la]tion shows that this product is

$$\begin{bmatrix} \cos 2\vartheta & \sin 2\vartheta \\ \sin 2\vartheta & -\cos 2\vartheta \end{bmatrix}.$$

1.19 Rotate the axes through ϑ, project onto the new x-axis, then rotate the ax[es] through $-\vartheta$. The required matrix is then

$$\begin{bmatrix} \cos\vartheta & -\sin\vartheta \\ \sin\vartheta & \cos\vartheta \end{bmatrix} \begin{bmatrix} 1 & 0 \\ 0 & 0 \end{bmatrix} \begin{bmatrix} \cos\vartheta & \sin\vartheta \\ -\sin\vartheta & \cos\vartheta \end{bmatrix}$$

$$= \begin{bmatrix} \cos\vartheta & 0 \\ \sin\vartheta & 0 \end{bmatrix} \begin{bmatrix} \cos\vartheta & \sin\vartheta \\ -\sin\vartheta & \cos\vartheta \end{bmatrix}$$

$$= \begin{bmatrix} \cos^2\vartheta & \sin\vartheta\cos\vartheta \\ \sin\vartheta\cos\vartheta & \sin^2\vartheta \end{bmatrix}.$$

1.20 The respective Hermite matrices are

$$\begin{bmatrix} 1 & 0 & \frac{1}{5} \\ 0 & 1 & \frac{7}{5} \\ 0 & 0 & 0 \end{bmatrix}; \quad \begin{bmatrix} 1 & 0 & 0 \\ 0 & 1 & 0 \\ 0 & 0 & 1 \end{bmatrix}; \quad \begin{bmatrix} 1 & 0 & 0 \\ 0 & 1 & 3 \\ 0 & 0 & 0 \end{bmatrix}.$$

1.21 A row-echelon form of the augmented matrix is

$$\begin{bmatrix} 1 & 2 & 0 & 3 & 1 \\ 0 & 1 & -1 & 1 & 0 \\ 0 & 0 & 1 & 0 & -2 \\ 0 & 0 & 0 & 0 & 1 \end{bmatrix}.$$

This can be obtained as follows (using the 'punning notation' ρ_i to mean row i):

$$\begin{bmatrix} 1 & 2 & 0 & 3 & 1 \\ 1 & 2 & 3 & 3 & 0 \\ 1 & 0 & 1 & 1 & 3 \\ 1 & 1 & 1 & 2 & 1 \end{bmatrix} \to \begin{bmatrix} 1 & 2 & 0 & 3 & 1 \\ 0 & 0 & 3 & 0 & -1 \\ 0 & -2 & 1 & -2 & 2 \\ 0 & -1 & 1 & -1 & 0 \end{bmatrix} \begin{matrix} \\ \rho_2 - \rho_1 \\ \rho_3 - \rho_1 \\ \rho_4 - \rho_1 \end{matrix}$$

$$\to \begin{bmatrix} 1 & 2 & 0 & 3 & 1 \\ 0 & 1 & -1 & 1 & 0 \\ 0 & -2 & 1 & -2 & 2 \\ 0 & 0 & 3 & 0 & -1 \end{bmatrix} \begin{matrix} \\ -\rho_4 \to \rho_2 \\ \\ \rho_2 \to \rho_4 \end{matrix} \to \begin{bmatrix} 1 & 2 & 0 & 3 & 1 \\ 0 & 1 & -1 & 1 & 0 \\ 0 & 0 & -1 & 0 & 2 \\ 0 & 0 & 3 & 0 & -1 \end{bmatrix} \begin{matrix} \\ \\ \rho_3 + 2\rho_2 \\ \\ \end{matrix}$$

$$\to \begin{bmatrix} 1 & 2 & 0 & 3 & 1 \\ 0 & 1 & -1 & 1 & 0 \\ 0 & 0 & 1 & 0 & -2 \\ 0 & 0 & 0 & 0 & -5 \end{bmatrix} \begin{matrix} \\ \\ -\rho_3 \\ \rho_4 + 3\rho_2 \end{matrix} \to \begin{bmatrix} 1 & 2 & 0 & 3 & 1 \\ 0 & 1 & -1 & 1 & 0 \\ 0 & 0 & 1 & 0 & -2 \\ 0 & 0 & 0 & 0 & 1 \end{bmatrix} \begin{matrix} \\ \\ \\ -\frac{1}{5}\rho_4 \end{matrix}.$$

It is clear from this that the coefficient matrix is of rank 3 whereas the augmented matrix is of rank 4. Consequently, the system has no solution.

1.22 A row-echelon form of the augmented matrix is

$$\begin{bmatrix} 1 & 1 & 0 & 1 & 4 \\ 0 & -2 & 0 & -6 & -1 \\ 0 & 0 & 1 & -1 & 1 \\ 0 & 0 & 0 & 0 & \lambda-1 \end{bmatrix}.$$

It is clear from this that the coefficient matrix and the augmented matri have the same rank (i.e. the system has a solution) only when $\lambda = 1$. In thi case, since the rank is 3 and the number of unknowns is 4, we can assig $4-3=1$ solution parameter. Taking this to be t, we have $z=1+$ $-2y = 6t-1, x+y = 4-t$ so that the general solution is

$$z = 1+t,$$
$$y = -3t + \tfrac{1}{2},$$
$$x = 2t + \tfrac{7}{2}.$$

1.23 We have, by the obvious row operations,

$$\begin{bmatrix} 1 & 1 & -1 & 0 & b \\ 2 & -1 & 1 & -3 & a \\ 4 & 1 & -1 & -3 & c \end{bmatrix} \to \begin{bmatrix} 1 & 1 & -1 & 0 & b \\ 0 & -3 & 3 & -3 & a-2b \\ 0 & -3 & 3 & -3 & c-4b \end{bmatrix}$$

from which we see that the system has integer solutions if and only $a-2b = c-4b$ and $a-2b$ is divisible by 3; i.e. $a+2b=c$ and 3 divid $a+b$.

1.24 A row-echelon form of the augmented matrix is

$$\begin{bmatrix} 1 & 1 & 1 & 1 & a \\ 0 & 1 & 1 & 0 & \frac{1}{2}(a-b) \\ 0 & 0 & 1 & 1 & \frac{1}{2}(a+c) \\ 0 & 0 & 0 & 0 & \frac{1}{4}(d+3a)-\frac{1}{2}(a-b)+\frac{1}{2}(c+a) \end{bmatrix}.$$

A solution exists if and only if the rank of the coefficient matrix is the sa as the rank of the augmented matrix. This is so if and only if $0 = \frac{1}{4}(d+3a)$ $\frac{1}{2}(a-b)+\frac{1}{2}(c+a)$, i.e. if and only if $3a+2b+2c+d=0$. Since $a,b,c,d >$ this condition is not met, so the system has no solution.

1.25 We have

$$\begin{bmatrix} 2 & 1 & 1 & -6\beta \\ \gamma & 3 & 2 & 2\beta \\ 2 & 1 & \gamma+1 & 4 \end{bmatrix} \to \begin{bmatrix} 2 & 1 & 1 & -6\beta \\ 0 & 6-\gamma & 4-\gamma & 4\beta+6\beta\gamma \\ 0 & 0 & \gamma & 4+6\beta \end{bmatrix} \begin{matrix} \\ 2\rho_2 - \\ \rho_3 - \rho \end{matrix}$$

If now $\gamma \neq 0$ and $\gamma \neq 6$ then we can reduce this matrix to Hermite form with rank 3, in which case a unique solution exists.

Suppose now that $\gamma = 0$. Then the reduced augmented matrix becomes

$$\begin{bmatrix} 2 & 1 & 1 & -6\beta \\ 0 & 6 & 4 & 4\beta \\ 0 & 0 & 0 & 4+6\beta \end{bmatrix}$$

from which we see that the system has a solution if and only if $0 = 4 + 6\beta$, i.e. if and only if $\beta = -\frac{2}{3}$. When this is so, the matrix becomes

$$\begin{bmatrix} 2 & 1 & 1 & 4 \\ 0 & 6 & 4 & -\frac{8}{3} \\ 0 & 0 & 0 & 0 \end{bmatrix} \to \begin{bmatrix} 1 & \frac{1}{2} & \frac{1}{2} & 2 \\ 0 & 1 & \frac{2}{3} & -\frac{4}{9} \\ 0 & 0 & 0 & 0 \end{bmatrix} \to \begin{bmatrix} 1 & 0 & \frac{1}{6} & \frac{20}{9} \\ 0 & 1 & \frac{2}{3} & -\frac{4}{9} \\ 0 & 0 & 0 & 0 \end{bmatrix}$$

so the general solution is

$$x = \tfrac{20}{9} - \tfrac{1}{6}z,$$
$$y = -\tfrac{4}{9} - \tfrac{2}{3}z,$$

where z is arbitrary.

Suppose now that $\gamma = 6$. Then the reduced augmented matrix becomes

$$\begin{bmatrix} 2 & 1 & 1 & -6\beta \\ 0 & 0 & -2 & 40\beta \\ 0 & 0 & 6 & 4+6\beta \end{bmatrix} \to \begin{bmatrix} 2 & 1 & 1 & -6\beta \\ 0 & 0 & -2 & 40\beta \\ 0 & 0 & 0 & 4+126\beta \end{bmatrix}$$

so a solution exists if and only if $\beta = -\frac{2}{63}$. In this case the matrix becomes

$$\begin{bmatrix} 2 & 1 & 1 & \frac{12}{63} \\ 0 & 0 & 1 & \frac{40}{63} \\ 0 & 0 & 0 & 0 \end{bmatrix} \to \begin{bmatrix} 2 & 1 & 0 & -\frac{28}{63} \\ 0 & 0 & 1 & \frac{40}{63} \\ 0 & 0 & 0 & 0 \end{bmatrix}$$

whence the general solution is

$$z = \tfrac{40}{63},$$
$$x = -\tfrac{2}{9} - \tfrac{1}{2}y,$$

where y is arbitrary.

1.26 A row-echelon form of the augmented matrix is

$$\begin{bmatrix} 1 & -1 & 0 & -1 & -5 & \alpha \\ 0 & 3 & -1 & -2 & 11 & \beta - 2\alpha \\ 0 & 0 & 5 & -5 & -25 & 3(\gamma-\alpha) - 2(\beta - 2\alpha) \\ 0 & 0 & 0 & 0 & 0 & p \end{bmatrix}$$

where $p = 5[3(\delta - \alpha) - 5(\beta - 2\alpha)] - 11[3(\gamma - \alpha) - 2(\beta - 2\alpha)]$. A solution exists, therefore, if and only if

$$\begin{aligned} 0 &= 15\delta - 15\alpha - 25\beta + 50\alpha - 33\gamma + 33\alpha + 22\beta - 44\alpha \\ &= 24\alpha - 3\beta - 33\gamma + 15\delta \end{aligned}$$

i.e. if and only if $8\alpha - \beta - 11\gamma + 5\delta = 0$.

When $\alpha = \beta = -1$, $\gamma = 3$ and $\delta = 8$ the augmented matrix reduces to the Hermite form

$$\begin{bmatrix} 1 & 0 & 0 & -2 & -3 & 0 \\ 0 & 1 & 0 & -1 & 2 & 1 \\ 0 & 0 & 1 & -1 & -5 & 2 \\ 0 & 0 & 0 & 0 & 0 & 0 \end{bmatrix}$$

and so the general solution is

$$\begin{aligned} x &= 2u + 3t, \\ y &= u - 2t + 1, \\ z &= u + 5t + 2, \end{aligned}$$

where u, t are arbitrary.

Solutions to Chapter 2

2.1 The respective inverses are

$$\begin{bmatrix} 1 & 1 & -3 \\ 0 & 1 & -1 \\ -1 & -2 & 5 \end{bmatrix}; \quad \begin{bmatrix} 4 & 1 & -3 \\ 1 & 1 & -1 \\ -11 & -4 & 9 \end{bmatrix}; \quad \begin{bmatrix} 1 & 0 & 0 \\ -\frac{1}{2} & \frac{1}{2} & 0 \\ 0 & -\frac{1}{3} & \frac{1}{3} \end{bmatrix}$$

For example, to obtain the inverse of

$$A = \begin{bmatrix} 3 & 1 & 2 \\ 1 & 2 & 1 \\ 1 & 1 & 1 \end{bmatrix},$$

consider

$$[A \mid I_n] = \left[\begin{array}{ccc|ccc} 3 & 1 & 2 & 1 & 0 & 0 \\ 1 & 2 & 1 & 0 & 1 & 0 \\ 1 & 1 & 1 & 0 & 0 & 1 \end{array}\right].$$

Reducing this to Hermite normal form, we obtain

$$\left[\begin{array}{ccc|ccc} 3 & 1 & 2 & 1 & 0 & 0 \\ 1 & 2 & 1 & 0 & 1 & 0 \\ 1 & 1 & 1 & 0 & 0 & 1 \end{array}\right]$$

$$\to \left[\begin{array}{ccc|ccc} 1 & -3 & 0 & 1 & -2 & 0 \\ 0 & 1 & 0 & 0 & 1 & -1 \\ 1 & 1 & 1 & 0 & 0 & 1 \end{array}\right] \begin{array}{l} \rho_1 - 2\rho_2 \\ \rho_2 - \rho_3 \\ \end{array}$$

$$\to \left[\begin{array}{ccc|ccc} 1 & 0 & 0 & 1 & 1 & -3 \\ 0 & 1 & 0 & 0 & 1 & -1 \\ 1 & 0 & 1 & 0 & -1 & 2 \end{array}\right] \begin{array}{l} \rho_1 + 3\rho_2 \\ \\ \rho_3 - \rho_2 \end{array}$$

$$\rightarrow \left[\begin{array}{ccc|ccc} 1 & 0 & 0 & 1 & 1 & -3 \\ 0 & 1 & 0 & 0 & 1 & -1 \\ 0 & 0 & 1 & -1 & -2 & 5 \end{array}\right] \quad \rho_3 - \rho_1$$

$$= [I_n \mid A^{-1}].$$

2.2 Expanding by the first row, we have

$$\det \begin{bmatrix} x & 2 & 0 & 3 \\ 1 & 2 & 3 & 3 \\ 1 & 0 & 1 & 1 \\ 1 & 1 & 1 & 3 \end{bmatrix}$$

$$= x \det \begin{bmatrix} 2 & 3 & 3 \\ 0 & 1 & 1 \\ 1 & 1 & 3 \end{bmatrix} - 2 \det \begin{bmatrix} 1 & 3 & 3 \\ 1 & 1 & 1 \\ 1 & 1 & 3 \end{bmatrix} - 3 \det \begin{bmatrix} 1 & 2 & 3 \\ 1 & 0 & 1 \\ 1 & 1 & 1 \end{bmatrix}$$

$$= 4x + 8 - 6$$

$$= 4x + 2$$

so the matrix is invertible for all $x \neq -\frac{1}{2}$.

2.3 Expanding by the first row, then by the first column of the resulting 5 × matrix, then by the first row of the resulting 4 × 4 matrix, and so on, we se that the determinant of the original matrix is $-acefhm$. Thus the matrix is invertible if and only if a, c, e, f, h, m are all non-zero. In this case we ca compute the inverse, using row operations in the usual way (i.e. proceedin from $A \mid I_6$ to $I_6 \mid A^{-1}$). The answer is

$$\begin{bmatrix} 0 & \frac{1}{f} & 0 & -\frac{b}{fh} & 0 & \frac{bd}{fhm} \\ \frac{1}{a} & 0 & 0 & 0 & 0 & 0 \\ 0 & 0 & 0 & \frac{1}{h} & 0 & -\frac{d}{hm} \\ -\frac{g}{ca} & 0 & \frac{1}{c} & 0 & 0 & 0 \\ 0 & 0 & 0 & 0 & 0 & \frac{1}{m} \\ \frac{kg}{eca} & 0 & -\frac{k}{ec} & 0 & \frac{1}{e} & 0 \end{bmatrix}.$$

2.4 Simple computations show that

$$P^{-1} = \tfrac{1}{9}\begin{bmatrix} -2 & 4 & 1 \\ 1 & -2 & 4 \\ 4 & 1 & -2 \end{bmatrix},$$

$$P^{-1}AP = \begin{bmatrix} 9a & 0 & 0 \\ 0 & 9b & 0 \\ 0 & 0 & 9c \end{bmatrix}.$$

It follows that $\det(P^{-1}AP) = 3^6abc$. Since

$$\begin{aligned}\det(P^{-1}AP) &= \det P^{-1} \det A \det P \\ &= \det A\,(\det P)^{-1} \det P \\ &= \det A,\end{aligned}$$

we conclude that $\det A = 3^6abc$.

2.5 Subtracting the first column from each of the others, then removing factors, and then repeating this process, we see that

$$\det \Delta = \begin{vmatrix} 1 & 0 & 0 & 0 \\ a & x-a & b-a & c-a \\ a^2 & x^2-a^2 & b^2-a^2 & c^2-a^2 \\ a^3 & x^3-a^3 & b^3-a^3 & c^3-a^3 \end{vmatrix}$$

$$= (x-a)(b-a)(c-a)\begin{vmatrix} 1 & 1 & 1 \\ x+a & b+a & c+a \\ x^2+ax+a^2 & b^2+ab+a^2 & c^2+ac+a^2 \end{vmatrix}$$

$$= (x-a)(b-a)(c-a)(b-x)(c-x)\begin{vmatrix} 1 & 1 \\ b+x+a & c+x+a \end{vmatrix}$$

$$= (x-a)(x-b)(x-c)(a-c)(c-a).$$

2.6 We have

$$\begin{vmatrix} x & a & a & a \\ a & x & a & a \\ a & a & x & a \\ a & a & a & x \end{vmatrix} = \begin{vmatrix} x & a & a & a \\ a-x & x-a & 0 & 0 \\ a-x & 0 & x-a & 0 \\ a-x & 0 & 0 & x-a \end{vmatrix}$$

$$= \begin{vmatrix} x+3a & a & a & a \\ 0 & x-a & 0 & 0 \\ 0 & 0 & x-a & 0 \\ 0 & 0 & 0 & x-a \end{vmatrix}$$

$$= (x+3a)(x-a)^3.$$

Thus the solutions of the given equation are $x = -3a$ and $x = a$.

2.7 We have

$$\det \begin{bmatrix} x & a & ? & ? \\ y^2 & y & a & ? \\ yz^2 & z^2 & z & a \\ yzt^2 & zt^2 & t^2 & t \end{bmatrix} = \det \begin{bmatrix} x-ay & a & ? & ? \\ 0 & y & a & ? \\ 0 & z^2 & z & a \\ 0 & zt^2 & t^2 & t \end{bmatrix}$$

$$= (x-ay)\det \begin{bmatrix} y-az & a & ? \\ 0 & z & a \\ 0 & t^2 & t \end{bmatrix}$$

$$= (x-ay)(y-az)(z-at)t.$$

2.8 Subtracting the first row of B_r from each of the other rows yields

$$\det B_r = \det \begin{bmatrix} b & b & b & \dots & b & b \\ a-b & 0 & 0 & \dots & 0 & 0 \\ -2b & a-b & 0 & \dots & 0 & 0 \\ -2b & -2b & a-b & \dots & 0 & 0 \\ \vdots & \vdots & \vdots & & \vdots & \vdots \\ -2b & -2b & -2b & \dots & a-b & 0 \end{bmatrix}.$$

Subtracting the last column from each of the other columns, then expanding by the first row, we clearly obtain

$$\det B_r = (-1)^{r+1} b(a-b)^{r-1}.$$

Adding the last column of A_r to the first column, we obtain

$$\det A_r = \det \begin{bmatrix} a+b & b & b & \dots & b & b \\ 0 & a & b & \dots & b & b \\ 0 & -b & a & \dots & b & b \\ 0 & -b & -b & \dots & b & b \\ \vdots & \vdots & \vdots & & \vdots & \vdots \\ a-b & -b & -b & \dots & -b & a \end{bmatrix}.$$

Now we expand by the first column to obtain

$$\begin{aligned}\det A_r &= (a+b)\det A_{r-1} + (-1)^{r+1}(a-b)\det B_{r-1}\\ &= (a+b)\det A_{r-1} + (-1)^{r+1}(a-b)(-1)^r b(a-b)^{r-2}\\ &= (a+b)\det A_{r-1} - b(a-b)^{r-1}.\end{aligned}$$

For the last part, we use induction. Clearly, we have $\det A_1 = a = \frac{1}{2}[(a+b)^1 + (a-b)^1]$. Suppose then that the result holds for all $r < n$. Then

$$\begin{aligned}\det A_n &= (a+b)\tfrac{1}{2}[(a+b)^{n-1} + (a-b)^{n-1}] - b(a-b)^{n-1}\\ &= \tfrac{1}{2}[(a+b)^n - (a+b)^{n-1}(a+b-2b)]\\ &= \tfrac{1}{2}[(a+b)^n + (a-b)^n].\end{aligned}$$

2.9 Expanding by the first row of A_{n+2} we have

$$\Delta_{n+2} = 2\cos\vartheta\,\Delta_{n+1} - \det\begin{bmatrix} 1 & 1 & 0 & \dots & 0\\ 0 & 2\cos\vartheta & 1 & \dots & 0\\ 0 & 1 & 2\cos\vartheta & \dots & 0\\ \vdots & \vdots & \vdots & & \vdots\\ 0 & 0 & 0 & \dots & 2\cos\vartheta\end{bmatrix}.$$

Expanding by the first column, we obtain clearly

$$\Delta_{n+2} = 2\cos\vartheta\,\Delta_{n+1} - \Delta_n$$

so that

$$\Delta_{n+2} - 2\cos\vartheta\,\Delta_{n+1} + \Delta_n = 0.$$

For the last part, we use induction. The result holds for $n=1$ since $\Delta_1 = 2\cos\vartheta = \sin 2\vartheta/\sin\vartheta$. For $n=2$ we have $\Delta_2 = 4\cos^2\vartheta - 1 = \sin 3\vartheta/\sin\vartheta$. Suppose then that the result holds for all $n < k$ with $k > 2$. Then

$$\begin{aligned}\sin\vartheta\,\Delta_k &= 2\cos\vartheta\sin k\vartheta - \sin(k-1)\vartheta\\ &= 2\cos\vartheta\sin k\vartheta - \sin(k\vartheta - \vartheta)\\ &= 2\cos\vartheta\sin k\vartheta - \cos\vartheta\sin k\vartheta + \cos k\vartheta\sin\vartheta\\ &= \cos\vartheta\sin k\vartheta + \cos k\vartheta\sin\vartheta\\ &= \sin(k+1)\vartheta,\end{aligned}$$

from which the result follows.

2.10 We have that

$$A_n = \begin{bmatrix} a_1+b_1 & b_1 & \dots & b_1 & b_1\\ b_2 & a_2+b_2 & \dots & b_2 & b_2\\ \vdots & \vdots & & \vdots & \vdots\\ b_{n-1} & b_{n-1} & \dots & a_{n-1}+b_{n-1} & b_{n-1}\\ b_n & b_n & \dots & b_n & b_n\end{bmatrix}.$$

Subtracting the last column of A_n from each of the others and then expanding by the first row, we obtain

$$\det A_n = b_n \prod_{i=1}^{n-1} a_i.$$

As for B_n, we have

$$B_n = \begin{bmatrix} a_1+b_1 & b_1 & \dots & b_1 \\ b_2 & a_2+b_2 & \dots & b_2 \\ \vdots & \vdots & & \vdots \\ b_n & b_n & \dots & a_n+b_n \end{bmatrix}.$$

Expanding by the last column we see that

$$\det B_n = \det A_n + a_n \det B_{n-1}.$$

For the last part, we use induction. The result clearly holds for $n = 1$ sinc $\det B_1 = a_1 + b_1$. Suppose then that the result holds for all $k < n$. Then w have

$$\begin{aligned}\det B_n &= b_n \Pi_{i=1}^{n-1} a_i + a_n(\Pi_{i=1}^{n-1} a_i + \Sigma_{i=1}^{n-1} (b_i \Pi_{j\neq i} a_j)) \\ &= \Pi_{i=1}^{n} a_i + \Sigma_{i=1}^{n} (b_i \Pi_{j\neq i} a_j).\end{aligned}$$

2.11 Suppose that $\det A = 0$. Then there exist $t_1, \dots, t_n$ with $\Sigma_{i=1}^{n} a_{ri}t_i = 0$. No let $|t_k| = \max_i |t_i|$. Then

$$a_{kk}t_k = -\sum_{i\neq k} a_{ki}t_i.$$

But then

$$|a_{kk}|\,|t_k| = \sum_{i\neq k} |a_{ki}|\,|t_i| \leqslant \sum_{i\neq k} |a_{ki}|\,|t_k|$$

from which it follows that we cannot have $|a_{kk}| > \Sigma_{i\neq k} |a_{ki}|$. Thus, if A dominated by its diagonal elements then we must have $\det A \neq 0$.

2.12 We have that

$$\begin{vmatrix} A & B \\ B & A \end{vmatrix} = \begin{vmatrix} A & B \\ B-A & A-B \end{vmatrix} = \begin{vmatrix} A+B & B \\ 0 & A-B \end{vmatrix}$$

$$= \left| \begin{bmatrix} I & B \\ 0 & A-B \end{bmatrix} \begin{bmatrix} A+B & 0 \\ 0 & I \end{bmatrix} \right|$$

$$= \begin{vmatrix} I & B \\ 0 & A-B \end{vmatrix} \begin{vmatrix} A+B & 0 \\ 0 & I \end{vmatrix}$$
$$= |A-B|\,|A+B|.$$

2.13 Clearly,

$$\begin{bmatrix} A & 0 \\ B & C \end{bmatrix} \begin{bmatrix} P & Q \\ R & S \end{bmatrix} = \begin{bmatrix} AP & AQ \\ BP+CR & BQ+CS \end{bmatrix}$$

and this is of the given form if and only if

$$A = P^{-1}, \quad BP + CR = 0, \quad BQ + CS = S - RP^{-1}Q.$$

Since $B = -CRP^{-1}$ we have $C(S - RP^{-1}Q) = S - RP^{-1}Q$, so we can choose $C = I$ and then $B = -RP^{-1}$. Now it is clear that

$$|N|\,|M| = \begin{vmatrix} I & P^{-1}Q \\ 0 & S-RP^{-1}Q \end{vmatrix} = \begin{vmatrix} I & 0 \\ 0 & S-RP^{-1}Q \end{vmatrix} = |S - RP^{-1}Q|,$$

i.e. $|P^{-1}|\,|M| = |S - RP^{-1}Q|$, so

$$|M| = |P|\,|S - RP^{-1}Q| = |PS - PRP^{-1}Q|.$$

If now $PR = RP$ then $PRP^{-1} = R$ and so $|M| = |PS - RQ|$. Likewise, we also have

$$|M| = |S - RP^{-1}Q|\,|P| = |SP - RP^{-1}QP|$$

so if $PQ = QP$ then $|M| = |SP - RQ|$.

Solutions to Chapter 3

3.1 Expanding along the first row, we have

$$\det\begin{bmatrix} 1-\lambda & -4 & 0 \\ 2 & -2-\lambda & -2 \\ -2.5 & 1 & -2-\lambda \end{bmatrix}$$
$$= (1-\lambda)[(2+\lambda)^2+2]+4(-4-2\lambda-5)$$
$$= (1-\lambda)(\lambda^2+4\lambda+6)-36-8\lambda$$
$$= \lambda^2+4\lambda+6-\lambda^3-4\lambda^2-6\lambda-36-8\lambda$$
$$= -\lambda^3-3\lambda^2-10\lambda-30,$$

so the characteristic polynomial is

$$\chi_A(X) = X^3+3X^2+10X+30 = (X+3)(X^2+10).$$

Since $A+3I \neq 0$ and $A^2+10I \neq 0$, this is also the minimum polynomi[al] $m_A(X)$.

The other matrices are treated in a similar way. The answers are

$$\chi_B(X) = (X-1)^3 = m_B(X);$$
$$\chi_C(X) = X^3-X^2-X+3 = m_C(X);$$
$$\chi_D(X) = X^3-X^2+X-2 = m_D(X).$$

3.2 Let $P = I_n - AB$ and $Q = I_n - BA$. Then we have

$$QBP^{-1}A = (I_n-BA)B(I_n-AB)^{-1}A$$
$$= (B-BAB)(I_n-AB)^{-1}A$$
$$= B(I_n-AB)(I_n-AB)^{-1}A$$
$$= BA$$

whence we see that

$$Q(I_n+BP^{-1}A) = Q+BA = I_n$$

and hence that Q^{-1} exists and is $I_n+BP^{-1}A$.

Suppose that λ is not an eigenvalue of AB. Then $\lambda I_n - AB$ is invertible. If $\lambda \neq 0$ then $I_n - (1/\sqrt{\lambda})A.(1/\sqrt{\lambda})B$ is invertible, whence so also is $I_n - (1/\sqrt{\lambda})B.(1/\sqrt{\lambda})A$, whence so also is $\lambda I_n - BA$, so that λ is not an eigenvalue of BA. If $\lambda = 0$ then AB is invertible, whence so is BA and so λ is not an eigenvalue of BA in this case also. We conclude that if λ is an eigenvalue of AB then it is an eigenvalue of BA, whence the result follows.

3.3 If $A = PQ$ and $B = QP$ then, using a simple inductive argument, we have that, for all positive integers i,

$$A^{i+1} = (PQ)^{i+1} = P(QP)^i Q = PB^i Q.$$

If now $h(X) = z_0 + z_1 X + \cdots + z_n X^n$ then we have

$$\begin{aligned} Ah(A) &= z_0 A + z_1 A^2 + \cdots + z_n A^{n+1} \\ &= z_0 PQ + z_1 PBQ + \cdots z_n PB^n Q \\ &= P(z_0 I_n + z_1 B + \cdots + z_n B^n)Q \\ &= Ph(B)Q. \end{aligned}$$

It follows immediately that $Am_B(A) = Pm_B(B)Q = 0$. Similarly we have that $Bm_A(B) = 0$. Consequently we have

$$m_A(X) \mid Xm_B(X) \quad \text{and} \quad m_B(X) \mid Xm_A(X)$$

and so can write $Xm_B(X) = p(X)m_A(X)$ and $Xm_A(X) = q(X)m_B(X)$. Comparing the degrees of each of these equations, we deduce that $\deg p + \deg q = 2$. Thus, either $\deg p = 0$, in which case $p(X) = 1$ and $Xm_B(X) = m_A(X)$, or $\deg q = 0$, in which case $q(X) = 1$ and $Xm_A(X) = m_B(X)$, or $\deg p = \deg q = 1$, in which case $m_A(X) = m_B(X)$.

We can write

$$A = \begin{bmatrix} 1 & 1 & \dots & 1 \\ 2 & 2 & \dots & 2 \\ \vdots & \vdots & & \vdots \\ r & r & \dots & r \end{bmatrix} = \begin{bmatrix} 1 \\ 2 \\ \vdots \\ r \end{bmatrix} [1 \quad 1 \quad \dots \quad 1] = PQ$$

where P is the column matrix and Q is the row matrix. Note that then $B = QP$ is the 1×1 matrix whose entry is $\frac{1}{2}r(r+1)$. Clearly, we have

$$m_B(X) = -\tfrac{1}{2}r(r+1) + X.$$

Clearly, $m_A(X) \neq m_B(X)$ and $m_B(X) \neq Xm_A(X)$. Thus we must have $m_A(X) = Xm_B(X) = -\frac{1}{2}X + X^2$.

3.4 We prove that $|[A^r]_{ij}| \leqslant k^r n^{r-1}$ by induction. The result clearly holds for

$r = 1$. Suppose that it holds for r. Then we have

$$\begin{aligned} |[A^{r+1}]_{ij}| &= |\Sigma_{t=1}^{n} a_{it}[A^r]_{tj}| \\ &\leqslant \Sigma_{t=1}^{n} |a_{it}|\,|[A^r]_{tj}| \\ &\leqslant \Sigma_{t=1}^{n} k\,.\,k^r n^{r-1} \\ &= k^{r+1}n^{r-1}\,\Sigma_{t=1}^{n} 1 \\ &= k^{r+1}n^r. \end{aligned}$$

(*a*) If $|\beta| < 1/nk$ then

$$\begin{aligned} &1 + |\beta|\,|[A]_{ij}| + |\beta^2|\,|[A^2]_{ij}| + \cdots + |\beta^r|\,|[A^r]_{ij}| + \cdots \\ &\leqslant 1 + k|\beta| + k^2 n|\beta|^2 + \cdots + k^r n^{r-1}|\beta|^r + \cdots \\ &= 1 + k|\beta|(1 + kn|\beta| + \cdots + k^{r-1}n^{r-1}|\beta|^{r-1} + \cdots) \end{aligned}$$

which, if $|\beta| \leqslant 1/nk$, is less than or equal to a geometric series which con- verges. Thus we see that if $|\beta| \leqslant 1/nk$ then $S_\beta(A)$ is absolutely convergent hence convergent.

(*b*) If $S_\beta(A)$ is convergent then $\lim_{t\to\infty}\beta^t A^t = 0$ so

$$\begin{aligned} &(I_n - \beta A)(I_n + \beta A + \beta^2 A^2 + \cdots) \\ &= \lim_{t\to\infty}(I_n - \beta A)(I_n + \beta A + \cdots + \beta^n A^n) \\ &= \lim_{t\to\infty}(I_n - \beta^{t+1}A^{t+1}) \\ &= I_n. \end{aligned}$$

Consequently $I_n - \beta A$ has an inverse which is the sum of the series.

For the last part, let λ be an eigenvalue of A. Then $\lambda I_n - A$ is not inve tible. Suppose, by way of obtaining a contradiction, that $|\lambda| > nk$. The $1/|\lambda| < 1/nk$. Consequently, if we let $\beta = 1/\lambda$ we have, by (*a*), that $S_\beta(A$ converges and so, by (*b*), $I_n - \beta A = I_n - (1/\lambda)A$ is invertible. It follows th $\lambda I_n - A$ is invertible, a contradiction. Hence we must have $|\lambda| \leqslant nk$.

3.5 We have

$$\begin{aligned} &\det\begin{bmatrix} \lambda+2 & 3 & 3 \\ 1 & \lambda & 1 \\ -5 & -5 & \lambda-6 \end{bmatrix} \\ &= (\lambda+2)(\lambda^2 - 6\lambda + 5) - 3(\lambda - 6 + 5) + 3(-5 + 5\lambda) \\ &= (\lambda+2)(\lambda-5)(\lambda-1) - 3(\lambda-1) + 15(\lambda-1) \\ &= (\lambda-1)[(\lambda+2)(\lambda-5) + 12] \\ &= (\lambda-1)(\lambda^2 - 3\lambda + 2) \\ &= (\lambda-1)^2(\lambda-2) \end{aligned}$$

and so the characteristic polynomial is $(X-1)^2(X-2)$. The eigenvalues are 1, 2 and the corresponding general eigenvectors are respectively

$$\begin{bmatrix} x \\ y \\ -x-y \end{bmatrix} \quad (x \neq 0 \text{ or } y \neq 0) \quad \text{and} \quad \begin{bmatrix} 3y \\ y \\ -5y \end{bmatrix} \quad (y \neq 0).$$

3.6 The characteristic polynomial of the first matrix is $X(X-1)(X-2)$ so the eigenvalues are 0, 1, 2. The corresponding general eigenvectors are respectively

$$\begin{bmatrix} a \\ 0 \\ -a \end{bmatrix} \ (a \neq 0); \quad \begin{bmatrix} 0 \\ b \\ 0 \end{bmatrix} \ (b \neq 0); \quad \begin{bmatrix} c \\ 0 \\ c \end{bmatrix} \ (c \neq 0).$$

The characteristic polynomial of the second matrix is $X(X-1)(X+1)$ so the eigenvalues are 0, 1, -1. The corresponding general eigenvectors are respectively

$$\begin{bmatrix} a \\ -a \\ a \end{bmatrix} \ (a \neq 0); \quad \begin{bmatrix} -b \\ -2b \\ b \end{bmatrix} \ (b \neq 0); \quad \begin{bmatrix} 2c \\ -c \\ c \end{bmatrix} \ (c \neq 0).$$

3.7 The eigenvalues of the first matrix are 1, 2, 3. The corresponding general eigenvectors are respectively

$$\begin{bmatrix} 3z \\ z \\ z \end{bmatrix} \ (z \neq 0); \quad \begin{bmatrix} z \\ 2z \\ z \end{bmatrix} \ (z \neq 0); \quad \begin{bmatrix} 2z \\ z \\ z \end{bmatrix} \ (z \neq 0).$$

A matrix T such that $T^{-1}AT$ is diagonal is then, for example,

$$\begin{bmatrix} 3 & 1 & 2 \\ 1 & 2 & 1 \\ 1 & 1 & 1 \end{bmatrix}.$$

The eigenvalues of the second matrix are $-1, 2, 1$. General corresponding eigenvectors are respectively

$$\begin{bmatrix} z \\ 0 \\ -z \end{bmatrix} \ (z \neq 0); \quad \begin{bmatrix} z \\ z \\ -2z \end{bmatrix} \ (z \neq 0); \quad \begin{bmatrix} -3z \\ -z \\ 5z \end{bmatrix} \ (z \neq 0).$$

A matrix T such that $T^{-1}AT$ is diagonal is then, for example,

$$\begin{bmatrix} 1 & 1 & -3 \\ 0 & 1 & -1 \\ -1 & -2 & 5 \end{bmatrix}.$$

3.8 The first part is straightforward. The inverse of P is

$$\begin{bmatrix} \frac{1}{2} & -\frac{1}{2}\mathrm{i} \\ -\frac{1}{2}\mathrm{i} & \frac{1}{2} \end{bmatrix}$$

and a simple computation gives

$$P^{-1}AP = \begin{bmatrix} e^{\mathrm{i}\vartheta} & 0 \\ 0 & e^{-\mathrm{i}\vartheta} \end{bmatrix}.$$

3.9 We have that

$$\det \begin{bmatrix} a-\lambda & b \\ c & d-\lambda \end{bmatrix} = \lambda^2 - (a+d)\lambda + ad - bc$$

and so the eigenvalues of A are

$$\begin{aligned} \lambda &= \tfrac{1}{2}\{(a+d) \pm \sqrt{[(a+d)^2 - 4(ad-bc)]}\} \\ &= \tfrac{1}{2}\{(a+d) \pm \sqrt{[(a-d)^2 + 4bc]}\}. \end{aligned}$$

(*a*) Since $b, c > 0$ and $(a-d)^2 \geqslant 0$ it follows that the eigenvalues are rea and distinct.

(*b*) $a + d > 0$ and $\sqrt{[(a-d)^2 + 4bc]} > 0$ and therefore

$$\tfrac{1}{2}\{(a+d) + \sqrt{[(a-d)^2 + 4bc]}\} > 0.$$

(*c*) Since $b, c > 0$ and $(a-d)^2 + 4bc > (a-d)^2$ we have that

$$t = \tfrac{1}{2}\{(a-d) - \sqrt{[(a-d^2 + 4bc]}\} < 0.$$

Note that $t = a - \mu$ where μ is the largest eigenvalue of A. Now for ever eigenvector $\begin{bmatrix} x \\ y \end{bmatrix}$ that corresponds to μ we have $(a-\mu)x + by = 0$. Sinc $t = a - \mu < 0$ and $b > 0$ there are infinitely many such x, y with $x, y > 0$.

3.10 The characteristic polynomial of the first matrix is readily seen to b $X(X^2 - 2)$. The eigenvalues are therefore $0, \sqrt{2}, -\sqrt{2}$. Corresponding to th eigenvalue 0 the general eigenvector is $\begin{bmatrix} -z \\ 0 \\ z \end{bmatrix}$ with $z \neq 0$, so we can take th normalised eigenvector

$$\begin{bmatrix} -\frac{1}{\sqrt{2}} \\ 0 \\ \frac{1}{\sqrt{2}} \end{bmatrix}.$$

Similarly, corresponding to the eigenvalues $\sqrt{2}, -\sqrt{2}$ we can compute the normalised eigenvectors

$$\begin{bmatrix} \frac{1}{2} \\ \frac{1}{\sqrt{2}} \\ \frac{1}{2} \end{bmatrix} \text{ and } \begin{bmatrix} \frac{1}{2} \\ -\frac{1}{\sqrt{2}} \\ \frac{1}{2} \end{bmatrix}.$$

A suitable orthogonal matrix is then

$$P = \begin{bmatrix} \frac{1}{2} & -\frac{1}{\sqrt{2}} & \frac{1}{2} \\ -\frac{1}{\sqrt{2}} & 0 & \frac{1}{\sqrt{2}} \\ \frac{1}{2} & \frac{1}{\sqrt{2}} & \frac{1}{2} \end{bmatrix}.$$

As for the second matrix, the characteristic polynomial is $(X-1)(X-2)$-$(X-4)$ so the eigenvalues are $1, 2, 4$. Corresponding to the eigenvalue 1 a general eigenvector is $\begin{bmatrix} 0 \\ 0 \\ z \end{bmatrix}$ with $z \neq 0$, so we can take the normalised eigenvector $\begin{bmatrix} 0 \\ 0 \\ 1 \end{bmatrix}$.

Similarly, corresponding to the eigenvalues $2, 4$ we can compute the normalised eigenvectors

$$\begin{bmatrix} \frac{1}{\sqrt{2}} \\ \frac{1}{\sqrt{2}} \\ 0 \end{bmatrix} \text{ and } \begin{bmatrix} -\frac{1}{\sqrt{2}} \\ \frac{1}{\sqrt{2}} \\ 0 \end{bmatrix}.$$

A suitable orthogonal matrix is then

$$P = \begin{bmatrix} 0 & \frac{1}{\sqrt{2}} & -\frac{1}{\sqrt{2}} \\ 0 & \frac{1}{\sqrt{2}} & \frac{1}{\sqrt{2}} \\ 1 & 0 & 0 \end{bmatrix}.$$

3.11 We have that

$$\det \begin{bmatrix} \alpha - \lambda & \beta \\ 1 & -\lambda \end{bmatrix} = \lambda^2 - \alpha\lambda - \beta$$

so $\lambda_1^2 = \alpha\lambda_1 + \beta$ and $\lambda_2^2 = \alpha\lambda_2 + \beta$. Now

$$AP = \begin{bmatrix} \alpha & \beta \\ 1 & 0 \end{bmatrix} \begin{bmatrix} \lambda_1 & \lambda_2 \\ 1 & 1 \end{bmatrix} = \begin{bmatrix} \alpha\lambda_1 + \beta & \alpha\lambda_2 + \beta \\ \lambda_1 & \lambda_2 \end{bmatrix},$$

$$PD = \begin{bmatrix} \lambda_1 & \lambda_2 \\ 1 & 1 \end{bmatrix} \begin{bmatrix} \lambda_1 & 0 \\ 0 & \lambda_2 \end{bmatrix} = \begin{bmatrix} \lambda_1^2 & \lambda_2^2 \\ \lambda_1 & \lambda_2 \end{bmatrix}$$

and consequently we have that $AP = PD$. Since $\det P = \lambda_1 - \lambda_2 \neq 0$ we ha that P is invertible and so it follows that $P^{-1}AP = D$.

It is immediate from this that $A = PDP^{-1}$. Suppose, by way of inductio that $A^r = PD^rP^{-1}$. Then

$$\begin{aligned} A^{r+1} = AA^r &= APD^rP^{-1} \\ &= PDP^{-1}PD^rP^{-1} \\ &= PD^{r+1}P^{-1}. \end{aligned}$$

The given system of recurrence relations can be written in the matrix form

$$U_{n+1} = AU_n.$$

A simple inductive argument shows that $U_n = A^{n-1}U_1$ for every positi integer n. We can therefore compute U_n using A^{n-1}. But from the above see that

$$A^{n-1} = PD^{n-1}P^{-1}$$

$$= \frac{1}{\lambda_1 - \lambda_2}\begin{bmatrix}\lambda_1 & \lambda_2\\ 1 & 1\end{bmatrix}\begin{bmatrix}\lambda_1^{n-1} & 0\\ 0 & \lambda_2^{n-1}\end{bmatrix}\begin{bmatrix}1 & -\lambda_2\\ -1 & \lambda_1\end{bmatrix}$$

$$= \frac{1}{\lambda_1 - \lambda_2}\begin{bmatrix}\lambda_1^n - \lambda_2^n & \lambda_1\lambda_2^n - \lambda_2\lambda_1^n\\ \lambda_1^{n-1} - \lambda_2^{n-1} & \lambda_1\lambda_2^{n-1} - \lambda_2\lambda_1^{n-1}\end{bmatrix}.$$

Since $\begin{bmatrix}u_n\\ v_n\end{bmatrix} = A^n\begin{bmatrix}u_1\\ v_1\end{bmatrix}$ and $v_1 = u_0$ it follows that

$$u_n = \frac{1}{\lambda_1 - \lambda_2}(\lambda_1^n - \lambda_2^n)u_1 + \frac{1}{\lambda_1 - \lambda_2}(\lambda_1\lambda_2^n - \lambda_2\lambda_1^n)u_0$$

$$= \frac{\lambda_1^n}{\lambda_1 - \lambda_2}(u_1 - \lambda_2 u_0) + \frac{\lambda_2^n}{\lambda_1 - \lambda_2}(\lambda_1 u_0 - u_1).$$

3.12 Expanding by the first row we have

$$\Delta_n = \Delta_{n-1} + 4\det\begin{bmatrix}5 & -4 & \dots & 0 & 0\\ 0 & 1 & \dots & 0 & 0\\ \vdots & \vdots & & \vdots & \vdots\\ 0 & 0 & \dots & 1 & -4\\ 0 & 0 & \dots & 5 & 1\end{bmatrix}$$

$$= \Delta_{n-1} + 20\Delta_{n-2}.$$

It follows that

$$r_n = \begin{bmatrix}\Delta_n\\ \Delta_{n-1}\end{bmatrix} = \begin{bmatrix}\Delta_{n-1} + 20\Delta_{n-2}\\ \Delta_{n-1}\end{bmatrix}$$

$$= \begin{bmatrix}1 & 20\\ 1 & 0\end{bmatrix}\begin{bmatrix}\Delta_{n-1}\\ \Delta_{n-2}\end{bmatrix}$$

$$= Ar_{n-1}.$$

Consequently,

$$r_n = Ar_{n-1} = A^2 r_{n-2} = \dots = A^{n-2}r_2.$$

The characteristic polynomial of A is $(X-5)(X+4) = 0$ so the eigenvalues are $-4, 5$. Corresponding to the eigenvalue -4 the general eigenvector is $\begin{bmatrix}4y\\ -y\end{bmatrix}$ where $y \neq 0$; corresponding to the eigenvalue 5 the general eigenvector is $\begin{bmatrix}5y\\ y\end{bmatrix}$ where $y \neq 0$. Thus if

$$P = \begin{bmatrix} 4 & 5 \\ -1 & 1 \end{bmatrix}$$

we have that $P^{-1}AP = \begin{bmatrix} -4 & 0 \\ 0 & 5 \end{bmatrix}$. Using the fact that

$$P^{-1} = \tfrac{1}{9} \begin{bmatrix} 1 & -5 \\ 1 & 4 \end{bmatrix}$$

we can compute

$$\begin{aligned} A^{n-2} &= P \begin{bmatrix} (-4)^{n-2} & 0 \\ 0 & 5^{n-2} \end{bmatrix} P^{-1} \\ &= \tfrac{1}{9} \begin{bmatrix} 5^{n-1} - (-4)^{n-1} & 5(-4)^{n-1} + 4.5^{n-1} \\ 5^{n-2} - (-4)^{n-2} & 5(-4)^{n-2} + 4.5^{n-2} \end{bmatrix}. \end{aligned}$$

Using the fact that

$$r_2 = \begin{bmatrix} \Delta_2 \\ \Delta_1 \end{bmatrix} = \begin{bmatrix} 21 \\ 1 \end{bmatrix}$$

we deduce from the equation $r_n = A^{n-2} r_2$ that

$$\begin{aligned} \Delta_n &= \tfrac{1}{9} [25.5^{n-1} + (-4)^{n-1}(-21 + 5)] \\ &= \tfrac{1}{9} [5^{n+1} - (-4)^{n+1}]. \end{aligned}$$

3.13 The characteristic polynomial of A is $X(X-1)(X-4)$. The eigenvalues are 0, 1, 4 and the corresponding eigenvectors are respectively

$$\begin{bmatrix} -a \\ a \\ -a \end{bmatrix} \ (a \neq 0), \qquad \begin{bmatrix} -2b \\ b \\ b \end{bmatrix} \ (b \neq 0) \qquad \text{and} \qquad \begin{bmatrix} c \\ c \\ c \end{bmatrix} \ (c \neq 0).$$

Thus the matrix

$$P = \begin{bmatrix} -1 & -2 & 1 \\ 1 & 1 & 1 \\ -1 & 1 & 1 \end{bmatrix}$$

is such that

$$P^{-1}AP = \begin{bmatrix} 0 & 0 & 0 \\ 0 & 1 & 0 \\ 0 & 0 & 4 \end{bmatrix}.$$

Now

$$P^{-1}=\tfrac{1}{6}\begin{bmatrix}0 & 3 & -3\\ -2 & 0 & 2\\ 2 & 3 & 1\end{bmatrix} \quad\text{and}\quad A^n=P\begin{bmatrix}0 & 0 & 0\\ 0 & 1 & 0\\ 0 & 0 & 4^n\end{bmatrix}P^{-1},$$

from which we compute

$$A^n=\tfrac{1}{6}\begin{bmatrix}4+2.4^n & 3.4^n & -4+4^n\\ -2+2.4^n & 3.4^n & 2+4^n\\ -2+2.4^n & 3.4^n & 2+4^n\end{bmatrix}.$$

3.14 The coefficient matrix $A=\begin{bmatrix}2 & 6\\ 6 & -3\end{bmatrix}$ has eigenvalues $-7, 6$. Clearly we have

$$\begin{bmatrix}x_n\\ y_n\end{bmatrix}=A^{n-1}\begin{bmatrix}x_1\\ y_1\end{bmatrix}=A^{n-1}\begin{bmatrix}0\\ -1\end{bmatrix}.$$

Computing A^{n-1} using the technique described in the previous question, we obtain

$$\begin{bmatrix}x_n\\ y_n\end{bmatrix}=\begin{bmatrix}\tfrac{1}{13}[6(-7)^{n-1}-6^n]\\ -\tfrac{1}{13}[9(-7)^{n-1}+4.6^{n-1}]\end{bmatrix}.$$

3.15 Diagonalising A in the usual way, we have $P^{-1}AP=D$ where

$$P=\begin{bmatrix}1 & 1 & 0\\ 1 & -1 & 1\\ 0 & 0 & 1\end{bmatrix},$$

$$P^{-1}=\begin{bmatrix}\tfrac{1}{2} & \tfrac{1}{2} & -1\\ \tfrac{1}{2} & -\tfrac{1}{2} & 0\\ 0 & 0 & 1\end{bmatrix} \quad\text{and}\quad D=\begin{bmatrix}4 & 0 & 0\\ 0 & 9 & 0\\ 0 & 0 & 4\end{bmatrix}.$$

Now the matrix

$$\bar{D}=\begin{bmatrix}2 & 0 & 0\\ 0 & 3 & 0\\ 0 & 0 & 2\end{bmatrix}$$

has the property that $\bar{D}^2=D$ and consequently the matrix $B=P\bar{D}P^{-1}$ has the property that $B^2=A$. A simple computation shows that

$$B=\begin{bmatrix}\tfrac{5}{2} & -\tfrac{1}{2} & \tfrac{1}{2}\\ -\tfrac{1}{2} & \tfrac{5}{2} & -\tfrac{1}{2}\\ 0 & 0 & 2\end{bmatrix}.$$

Solutions to Chapter 4

4.1 Given $x = (x_1, \ldots, x_n) \in \mathrm{IR}^n$, the equation

$$x = \lambda_1 a_1 + \lambda_2 a_2 + \cdots + \lambda_n a_n$$

is equivalent to the system

$$\begin{aligned} x_1 &= \lambda_1 a_{11} + \lambda_2 a_{21} + \cdots + \lambda_n a_{n1} \\ x_2 &= \lambda_1 a_{12} + \lambda_2 a_{22} + \cdots + \lambda_n a_{n2} \\ &\vdots \\ x_n &= \lambda_1 a_{1n} + \lambda_2 a_{2n} + \cdots + \lambda_n a_{nn} \end{aligned}$$

i.e. to the system

$$A^{\mathrm{t}} \begin{bmatrix} \lambda_1 \\ \vdots \\ \lambda_n \end{bmatrix} = \begin{bmatrix} x_1 \\ \vdots \\ x_n \end{bmatrix}$$

where $A = [a_{ij}]_{n \times n}$. This matrix equation has a unique solution if and only A^{t} is invertible (i.e. has full rank n), and this is so if and only if A is invertib

4.2 Use the result of the previous exercise.

(*a*) $\{(1, 1, 1), (1, 2, 3), (2, -1, 1)\}$ is a basis since

$$A = \begin{bmatrix} 1 & 1 & 1 \\ 1 & 2 & 3 \\ 2 & -1 & 1 \end{bmatrix} \to \begin{bmatrix} 1 & 1 & 1 \\ 0 & 1 & 2 \\ 0 & -3 & -1 \end{bmatrix} \to \begin{bmatrix} 1 & 1 & 1 \\ 0 & 1 & 2 \\ 0 & 0 & 5 \end{bmatrix}$$

shows that A has Hermite form I_3, so is invertible.

(*b*) $\{(1, 1, 2), (1, 2, 5), (5, 3, 4)\}$ is not a basis since

$$A = \begin{bmatrix} 1 & 1 & 2 \\ 1 & 2 & 5 \\ 5 & 3 & 4 \end{bmatrix} \to \begin{bmatrix} 1 & 1 & 2 \\ 0 & 1 & 3 \\ 0 & -2 & -6 \end{bmatrix} \to \begin{bmatrix} 1 & 1 & 2 \\ 0 & 1 & 3 \\ 0 & 0 & 0 \end{bmatrix}$$

shows that the Hermite form of A is not I_3, so A is not invertible.

4.3 The matrix

$$A = \begin{bmatrix} 1 & 1 & 0 & 0 \\ -1 & -1 & 1 & 2 \\ 1 & -1 & 1 & 3 \\ 0 & 1 & -1 & -3 \end{bmatrix}$$

has Hermite form I_4, so is invertible, so the given set constitutes a basis of $\mathbb{R}^4$. We now have to find $\alpha, \beta, \gamma, \delta$ such that

$$A^t \begin{bmatrix} \alpha \\ \beta \\ \gamma \\ \delta \end{bmatrix} = \begin{bmatrix} a \\ b \\ c \\ d \end{bmatrix}.$$

Working in the usual way with the augmented matrix, we have

$$\begin{bmatrix} 1 & -1 & 1 & 0 & a \\ 0 & 0 & -2 & 1 & b-a \\ 0 & 1 & 1 & -1 & c \\ 0 & 2 & 3 & -3 & d \end{bmatrix} \to$$

$$\begin{bmatrix} 1 & -1 & 1 & 0 & a \\ 0 & 1 & 1 & -1 & c \\ 0 & 0 & 0 & -1 & b-a+2d-4c \\ 0 & 0 & 1 & -1 & d-2c \end{bmatrix}$$

whence we see that $\alpha = b + c$, $\beta = 3c - d$, $\gamma = -b + a - d + 2c$, $\delta = -b + a - 2d + 4c$.

4.4 The set

$$\{(1,-1,1,-1), (1,1,-1,1), (1,0,0,0), (0,1,0,0), (0,0,1,0), (0,0,0,1)\}$$

spans $\mathbb{R}^4$. To find a basis, we reject any vectors that are linear combinations of their predecessors in this list. Now

$$(1,0,0,0)=a(1,-1,1,-1)+b(1,1,-1,1)$$

gives $a=b=\frac{1}{2}$ so we can reject $(1,0,0,0)$. Next

$$(0,1,0,0)=a(1,-1,1,-1)+b(1,1,-1,1)$$

gives $a+b=0$, $a-b=0$, $-a+b=1$ which are inconsistent; so we retair $(0,1,0,0)$. Next

$$(0,0,1,0)=a(1,-1,1,-1)+b(1,1,-1,1)+c(0,1,0,0)$$

gives $a+b=0$, $-a+b+c=0$, $a-b=1$, $-a+b=0$ which are inconsis tend; so we retain $(0,0,1,0)$. Since $\mathbb{R}^4$ has dimension 4, the linearly indepen dent set

$$\{(1,-1,1,-1),(1,1,-1,1),(0,1,0,0),(0,0,1,0)\}$$

is then a basis.

4.5 (*a*) False. Clearly, $(0,1)$ belongs to $\{(x,y)\in\mathbb{R}^2 \mid x_1<x_2\}$ but $(-1)(0,1$ does not.

(*b*) False. Take $\boldsymbol{b}=-\boldsymbol{a}_1$ or $\boldsymbol{a}_2-\boldsymbol{a}_1$, etc.

(*c*) False. Consider, for example, $\{(1,1),(2,2),(3,3),(4,4)\}$ in $\mathbb{R}^2$.

(*d*) True. Every spanning set contains a basis.

(*e*) False. The subspace $\{(x,x,x)\mid x\in\mathbb{R}\}$ has dimension 1 with basi $\{(1,1,1)\}$.

(*f*) True. We have $a(1,2,1)+b(2,2,1)=(a+2b,2a+2b,a+b)$ so w can take $a+b=y, b=x$.

(*g*) True. Proceed as in (*f*) but this time take $a+2b=2x, a=2y$.

(*h*) True. We can extend a basis for P to a basis for Q.

(*i*) False. For example, take $P=\{(x,x,x)\mid x\in\mathbb{R}\}$ and $Q=\{(x,y,0)$ $x,y\in\mathbb{R}\}$. We have $\dim P=1$, $\dim Q=2$ but $P\not\subseteq Q$.

(*j*) True. If U is an n-dimensional subspace of $\mathbb{R}^n$ with $U\neq\mathbb{R}^n$ the there exists $\boldsymbol{x}\in\mathbb{R}^n$ with $\boldsymbol{x}\notin U$. Add $\boldsymbol{x}$ to a basis of U to obtain a subset $\mathbb{R}^n$ containing $n+1$ linearly independent vectors, a contradiction. Hen $U=\mathbb{R}^n$.

4.6 (*a*) Yes. Clearly $U\neq\emptyset$ since $0\in U$, and U is closed under addition a multiplication by scalars.

(*b*) No. For example, $(1,0,0,0)\in U$ and $(0,1,0,0)\in U$ but their su does not belong to U.

(*c*) Yes. If $a^2+b^2=0$ then $a=b=0$ so that

$$U=\{(0,0,c,d)\mid c,d\in\mathbb{R}\}.$$

(*d*) No. For example, $(1,0,0,0) \in U$ and $(0,1,0,0) \in U$ but their sum does not belong to U.

(*e*) Yes. It is readily seen that U is non-empty and is closed under addition and multiplication by scalars.

(*f*) Yes. Same as (*e*).

4.7 The standard criteria are

(1) $x, y \in U \Rightarrow x + y \in U$;

(2) $x \in U, \lambda \in \mathbb{R} \Rightarrow \lambda x \in U$.

All of (*a*), (*b*), (*c*), (*d*) are correct, for the following reasons:

(*a*) Clearly, (1) and (2) imply (*a*). Conversely, assume (*a*). Take $a = b = 0$ to get $0 \in U$. Then (1) follows on taking $a = b = 1$, while (2) follows on taking $y = 0$.

(*b*) Clearly, (1) and (2) imply (*b*). Conversely, assume (*b*). Take $x = y$ and $a = -1$ to get $0 \in U$; then $a = 1$ gives (1) and $y = 0$ gives (2).

(*c*) Clearly, (1) and (2) imply (*c*). Conversely, assume (*c*). Take $a = 0$ to get $0 \in U$; then $a = 1$ gives (1) and $y = 0$ gives (2).

(*d*) Clearly, (1) and (2) imply (*d*). Conversely, assume (*d*). Take $a = 0$ to get $0 \in U$. Now for every $y \in U$ we have $-y \in U$ (take $x = 0, a = 1$). Consequently $ax + ay \in U$ and the result follows from (*c*).

4.8 If $\sum_{i=1}^{k} a_i = 1$ then we have

$$a_1(v - v_1) + \cdots + a_k(v - v_k) = v - v = 0$$

so that $v - v_1, \ldots, v - v_k$ are linearly dependent.

Conversely, suppose that $\sum_{i=1}^{k} a_i \neq 1$. Then from

$$\sum_{i=1}^{k} a_i(v - v_i) = \left(\sum_{i=1}^{k} a_i\right) v - v = \left(\sum_{i=1}^{k} a_i - 1\right) v$$

we see that v belongs to the subspace spanned by $\{v - v_1, \ldots, v - v_k\}$, whence so do $v_1, \ldots, v_k$. It follows that the subspace $\langle v - v_1, \ldots, v - v_k \rangle$ has dimension k and hence that $v - v_1, \ldots, v - v_k$ are linearly independent.

4.9 Choose a basis $\{v_1, \ldots, v_d\}$ of $X \cap Y$. Extend this to a basis

$$\{v_1, \ldots, v_d, v_{d+1}, \ldots, v_8\}$$

of X, and extend it also to a basis

$$\{v_1, \ldots, v_d, v'_{d+1}, \ldots, v'_9\}$$

of Y. Then each of $v'_{d+1}, \ldots, v'_9 \notin X$ (otherwise they are in $X \cap Y$), so

$$\{v_1, \ldots, v_d, v_{d+1}, \ldots, v_8, v'_{d+1}, \ldots, v'_9\}$$

is linearly independent, and therefore contains at most 10 elements. For this, we must have $d \geqslant 7$.

To see that this lower bound of 7 is attainable, consider $V = \mathbb{R}^{10}$ and take for X the subspace of those 10-tuples whose first and third components are 0, and for Y the subspace of those 10-tuples whose second components are 0.

4.10 We can have any first row other than a zero row, so there are $2^n - 1$ possible first rows. We can have any second row except zero and the first row, so there are $2^n - 2$ possible second rows. In general, there are $2^n - 2^{i-1}$ possible ith rows which are independent of the preceding rows. Hence there are

$$(2^n - 1)(2^n - 2)(2^n - 2^2) \cdots (2^n - 2^{n-1})$$

non-singular $n \times n$ matrices whose entries are 0 or 1. This product can be written in the form

$$2^{1+2+\cdots+(n-1)}(2^n - 1)(2^{n-1} - 1) \cdots (2^2 - 1)$$

which is the required form.

4.11 (*a*) If $A_1, \ldots, A_k$ are linearly independent then

$$\lambda_1 A_1 + \cdots + \lambda_k A_k = 0 \Rightarrow \lambda_1 = \cdots = \lambda_k = 0.$$

Suppose that

$$\lambda_1 X A_1 Y + \cdots \lambda_k X A_k Y = 0.$$

Multiplying by X^{-1} on the left and by Y^{-1} on the right, we obtain $\lambda_1 A_1 + \cdots + \lambda_k A_k = 0$ whence $\lambda_1 = \cdots = \lambda_k = 0$ and so $XA_1Y, \ldots, XA_kY$ are linearly independent. The converse is similar.

(*b*) If $\lambda_1 A_1 + \cdots + \lambda_k A_k = 0$ then $\lambda_1 A_1 B + \cdots + \lambda_k A_k B = 0$ whence $\lambda_1 = \cdots = \lambda_k = 0$. To show that the converse is false, take $B = 0$.

4.12 (*a*) The following matrices constitute a basis:

$$\begin{bmatrix} 1 & 0 \\ 0 & 0 \end{bmatrix}, \begin{bmatrix} 1 & 1 \\ 0 & 0 \end{bmatrix}, \begin{bmatrix} 0 & 0 \\ 1 & 1 \end{bmatrix}, \begin{bmatrix} 0 & 0 \\ 0 & 1 \end{bmatrix}.$$

(*b*) The following matrices constitute a basis:

$$\begin{bmatrix} 1 & 0 \\ 0 & 1 \end{bmatrix}, \begin{bmatrix} 1 & 1 \\ 0 & 1 \end{bmatrix}, \begin{bmatrix} 1 & 0 \\ 1 & 1 \end{bmatrix}, \begin{bmatrix} 1 & 1 \\ 1 & 0 \end{bmatrix}.$$

(*c*) The following matrices constitute a basis:

$$\begin{bmatrix} 1 & 0 \\ 0 & 1 \end{bmatrix}, \begin{bmatrix} 1 & 1 \\ 0 & 1 \end{bmatrix}, \begin{bmatrix} 1 & 0 \\ 1 & 1 \end{bmatrix}, \begin{bmatrix} 1 & -1 \\ 1 & 0 \end{bmatrix}.$$

Suppose that $\{A_1, A_2, A_3, A_4\}$ is a basis with $A_iA_j = A_jA_i$ for all i, j. Then there exist $\lambda_1, \lambda_2, \lambda_3, \lambda_4 \in \mathbb{Q}$ such that

$$\begin{bmatrix} 1 & 1 \\ 0 & 1 \end{bmatrix} = \lambda_1 A_1 + \lambda_2 A_2 + \lambda_3 A_3 + \lambda_4 A_4$$

and $\mu_1, \mu_2, \mu_3, \mu_4 \in \mathbb{Q}$ such that

$$\begin{bmatrix} 1 & 0 \\ 1 & 1 \end{bmatrix} = \mu_1 A_1 + \mu_2 A_2 + \mu_3 A_3 + \mu_4 A_4.$$

But we have that $(\Sigma \lambda_i A_i)(\Sigma \mu_i A_i) = (\Sigma \mu_i A_i)(\Sigma \lambda_i A_i)$ which contradicts the fact that

$$\begin{bmatrix} 1 & 1 \\ 0 & 1 \end{bmatrix}\begin{bmatrix} 1 & 0 \\ 1 & 1 \end{bmatrix} \neq \begin{bmatrix} 1 & 0 \\ 1 & 1 \end{bmatrix}\begin{bmatrix} 1 & 1 \\ 0 & 1 \end{bmatrix}.$$

The same argument shows that it is impossible to find such a basis for any $n \geqslant 2$. Hence $n = 1$ is the only possible value.

4.13 It is clear that $\langle a_1, a_2 \rangle \subseteq \langle b_1, b_2 \rangle$. Suppose first that $\det \begin{bmatrix} \alpha_1 & \alpha_2 \\ \beta_1 & \beta_2 \end{bmatrix} \neq 0$. Then $A = \begin{bmatrix} \alpha_1 & \alpha_2 \\ \beta_1 & \beta_2 \end{bmatrix}$ is invertible and we can rewrite

$$\begin{bmatrix} a_1 \\ a_2 \end{bmatrix} = A \begin{bmatrix} b_1 \\ b_2 \end{bmatrix}$$

in the form

$$\begin{bmatrix} b_1 \\ b_2 \end{bmatrix} = A^{-1} \begin{bmatrix} a_1 \\ a_2 \end{bmatrix}$$

whence $\langle b_1, b_2 \rangle \subseteq \langle a_1, a_2 \rangle$ and we have equality.

Conversely, suppose that $\langle a_1, a_2 \rangle = \langle b_1, b_2 \rangle$. Then $b_1, b_2 \in \langle a_1, a_2 \rangle$ gives $b_1 = \lambda_1 a_1 + \lambda_2 a_2$, $b_2 = \mu_1 a_1 + \mu_2 a_2$ so that

$$\begin{bmatrix} b_1 \\ b_2 \end{bmatrix} = \begin{bmatrix} \lambda_1 & \lambda_2 \\ \mu_1 & \mu_2 \end{bmatrix}\begin{bmatrix} a_1 \\ a_2 \end{bmatrix} = \begin{bmatrix} \lambda_1 & \lambda_2 \\ \mu_1 & \mu_2 \end{bmatrix}\begin{bmatrix} \alpha_1 & \alpha_2 \\ \beta_1 & \beta_2 \end{bmatrix}\begin{bmatrix} b_1 \\ b_2 \end{bmatrix}.$$

Since b_1, b_2 are linearly independent, this yields

$$\begin{bmatrix} \lambda_1 & \lambda_2 \\ \mu_1 & \mu_2 \end{bmatrix}\begin{bmatrix} \alpha_1 & \alpha_2 \\ \beta_1 & \beta_2 \end{bmatrix} = I_2$$

whence $\det \begin{bmatrix} \alpha_1 & \alpha_2 \\ \beta_1 & \beta_2 \end{bmatrix} \neq 0$.

4.14 The equation

$$x(2,2,1,3)+y(7,5,5,5)+z(3,2,2,1)+t(2,1,2,1)$$
$$=(6+\lambda,1+\lambda,-1+\lambda,2+\lambda)$$

is equivalent to

$$\begin{bmatrix}2&7&3&2\\2&5&2&1\\1&5&2&2\\3&5&1&1\end{bmatrix}\begin{bmatrix}x\\y\\z\\t\end{bmatrix}=\begin{bmatrix}6+\lambda\\1+\lambda\\-1+\lambda\\2+\lambda\end{bmatrix}$$

and, by the usual reduction process, we have

$$\begin{bmatrix}1&5&2&2&-1+\lambda\\0&3&1&2&-8+\lambda\\0&5&2&3&-3+\lambda\\0&10&5&5&-5+2\lambda\end{bmatrix}\rightarrow\begin{bmatrix}1&5&2&2&-1+\lambda\\0&1&0&1&-13+\lambda\\0&0&2&-2&62-4\lambda\\0&0&5&-5&125-8\lambda\end{bmatrix}$$

$$\rightarrow\begin{bmatrix}1&5&2&2&-1+\lambda\\0&1&0&1&-13+\lambda\\0&0&1&-1&31-2\lambda\\0&0&0&0&60-4\lambda\end{bmatrix}.$$

Thus we see that $x \in U$ if and only if $\lambda = 15$. For this value of λ we have th (non-unique) linear combination

$$x=(t+2)(2,2,1,3)+(2-t)(7,5,5,5)+(1+t)(3,2,2,1)$$
$$+t(2,1,2,1).$$

By the above reduction, we see that

$$\{(2,2,1,3),(7,5,5,5),(3,2,2,1)\}$$

is a basis for U.

If $\lambda \neq 15$ then $x \notin U$ and so, taking $\lambda = 0$, we have that

$$\{(2,2,1,3),(7,5,5,5),(3,2,2,1),(6,1,-1,2)\}$$

is a basis for $\mathbb{R}^4$.

4.15 It suffices to show that V is a subspace of the $\mathbb{Q}$-vector space $\mathbb{Q}$. That this so follows immediately from the fact that V is closed under addition a multiplication by scalars.

4.16 (*a*) W_1 is not a subspace. For, if f is defined by $f(x) = \frac{1}{2}$ for every $x \in$ then $f \in W$. But $\sqrt{2}f \notin W$ since $\sqrt{2}f(\frac{1}{2}) = 1/\sqrt{2} \notin \mathbb{Q}$.

(*b*) W_2 is a subspace. For if $\alpha, \beta \in \mathrm{IR}$ and $f, g \in W_2$ then

$$(\alpha f + \beta g)(\tfrac{1}{2}) = \alpha f(\tfrac{1}{2}) + \beta g(\tfrac{1}{2}) = \alpha f(1) + \beta g(1) = (\alpha f + \beta g)(1)$$

so that $\alpha f + \beta g \in W_2$.

(*c*) W_3 is a subspace. For if $\alpha, \beta \in \mathrm{IR}$ and $f, g \in W_3$ then

$$(\alpha f + \beta g)(\tfrac{1}{2}) = \alpha f(\tfrac{1}{2}) + \beta g(\tfrac{1}{2}) = \alpha . 0 + \beta . 0 = 0$$

so that $\alpha f + \beta g \in W_3$.

(*d*) W_4 is not a subspace. For if $f(x) = x^2$ for every $x \in \mathrm{IR}$ then $f \in W_4$ since $\mathrm{D}f(\tfrac{1}{2}) = 1$. But $2f \notin W_4$ since $2\mathrm{D}f(\tfrac{1}{2}) = 2$.

$W_2 \cap W_3 = \{f \in V \mid f(\tfrac{1}{2}) = 0, f(1) = 0\}$.

4.17 A_1, A_2, A_3 are linearly dependent since

$$(1,2,0) + (2,1,0) = (0,0,0),$$
$$(1,0,1) + 2(1,0,0) + 2(0,0,1) = (0,0,0),$$
$$(1,2,0) + (1,1,1) + 2(2,0,1) = (0,0,0).$$

A_4 is linearly independent since, in $\mathbb{Z}_3$,

$$\det\begin{bmatrix} 1 & 0 & 1 \\ 1 & 1 & 0 \\ 0 & 1 & 1 \end{bmatrix} = 1 + 1 = 2 \neq 0.$$

4.18 In order to show that the given set is a basis for $\mathbb{C}^3$ it suffices to show that the matrix

$$\begin{bmatrix} 3-\mathrm{i} & 2 & 1-\mathrm{i} \\ 2+2\mathrm{i} & 2+4\mathrm{i} & -2\mathrm{i} \\ 4 & 3 & -1 \end{bmatrix}$$

is invertible. Now for $\alpha, \beta, \gamma \in \mathbb{C}$ we have

$$(\alpha, \beta, \gamma) = \lambda(3-\mathrm{i}, 2+2\mathrm{i}, 4) + \mu(2, 2+4\mathrm{i}, 3) + \nu(1-\mathrm{i}, -2\mathrm{i}, -1).$$

Solving the resulting equations for λ, μ, ν we obtain

$$\begin{bmatrix} \lambda \\ \mu \\ \nu \end{bmatrix} = -\frac{1}{12+12\mathrm{i}}\begin{bmatrix} -2+2\mathrm{i} & 5-3\mathrm{i} & -6-6\mathrm{i} \\ 2-6\mathrm{i} & -7+5\mathrm{i} & 6+6\mathrm{i} \\ -2-10\mathrm{i} & -1+3\mathrm{i} & 6+6\mathrm{i} \end{bmatrix}\begin{bmatrix} \alpha \\ \beta \\ \gamma \end{bmatrix}.$$

This shows that the matrix is invertible. Moreover, putting $\alpha = 1$, $\beta = 0$, $\gamma = 0$ gives $(1, 0, 0)$ in terms of the basis; and similarly for the others.

4.19 $\mathbb{Z}_2^3$ has eight elements since each is of the form (a, b, c) and there are two choices for each of a, b, c.

Every basis of $\mathbb{Z}_2^3$ has three elements. Now any of the seven non-zero elements gives a linearly independent singleton set. There are six elements that are not multiples of this first choice; then four elements that are not linear combinations of these first two. It would appear, then, that there are $7 \times 6 \times 4$ possible bases. But the order in which we choose the basis elements is irrelevant, so we have counted each basis 3! times. Hence the number of bases is $(7 \times 6 \times 4)/6 = 28$.

$A = \{x_1 + x_2, x_2 + x_3, x_3 + x_1\}$ is linearly independent and therefore is a basis for V. Likewise, $B = \{x_1, x_1 + x_2, x_1 + x_2 + x_3\}$ is linearly independent and therefore is a basis.

If V is replaced by $\mathbb{Z}_2^3$ then the first of these statements is no longer true. For example,

$$\{(1, 0, 0), (0, 1, 0), (0, 0, 1)\}$$

is a basis for $\mathbb{Z}_2^3$ but

$$\{(1, 1, 0), (0, 1, 1), (1, 0, 1)\}$$

is not a basis since in this vector space we have

$$(1, 1, 0) + (0, 1, 1) + (1, 0, 1) = (0, 0, 0).$$

The second statement is true, however, since

$$\begin{aligned} 0 &= ax_1 + b(x_1 + x_2) + c(x_1 + x_2 + x_3) \\ &= (a + b + c)x_1 + (b + c)x_2 + cx_3 \end{aligned}$$

if and only if $a + b + c = 0$, $b + c = 0$, $c = 0$ which is the case if and only $a = b = c = 0$.

4.20 It is readily verified that E_n is closed under addition and multiplication by scalars, whence it is a subspace of the vector space Map($\mathbb{R}$, $\mathbb{R}$).

Suppose now that f is the zero map in E_1. Then we have

$$(\forall x \in \mathbb{R}) \quad a_0 + a_1 \cos x + b_1 \sin x = 0.$$

Taking $x = 0$ we obtain $a_0 + a_1 = 0$, and taking $x = \pi/2$ we obtain $a_0 + b_1 = 0$. Thus we have $a_1 = b_1 = -a_0$. Taking $x = \pi/4$ we obtain $a_0 + (1/\sqrt{2})a_1 + (1/\sqrt{2})b_1 = 0$ whence $a_0 = a_1 = b_1 = 0$. Suppose now, by way of induction, that the zero map of E_{n-1} (with $n \geqslant 2$) has all its coefficients zero and let f be the zero map of E_n. It is easily verified that $D^2 f + n^2 f$ is given by the prescription

$$(D^2 f + n^2 f)(x) = n^2 a_0 + \sum_{k=1}^{n-1} (n^2 - k^2)(a_k \cos kx + b_k \sin kx)$$

and since f is the zero map of E_n we see that D^2f+n^2f is the zero map of E_{n-1}. By the inductive hypothesis we therefore have

$$a_0, a_1, \ldots, a_{n-1}, b_1, \ldots, b_{n-1}$$

all zero and the formula for f reduces to

$$(\forall x \in \mathbb{R}) \quad 0 = f(x) = a_n \cos nx + b_n \sin nx.$$

Taking $x = 0$ we obtain $a_n = 0$; and taking $x = \pi/2n$ we obtain $b_n = 0$. Thus all the coefficients of f are zero and the result follows by induction.

It is clear that the $2n+1$ functions generate E_n. Moreover, by what we have just proved, the only linear combination of these $2n+1$ functions which is zero is the trivial linear combination. Hence these functions constitute a basis for E_n.

4.21 It is clear that the sum of two functions of the given form is also of that form; and that any scalar multiple of a function of that form is also of that form. The given set, E say, of such functions is therefore a subspace of the real vector space $\text{Map}(\mathbb{R} \setminus \{\alpha, \beta\}, \mathbb{R})$. That E is of dimension $r+s$ is immediate from the observation that every $f \in E$ can be written uniquely in the form $f = a_0 f_0 + \cdots + a_{r+s-1} f_{r+s-1}$ where, for $i = 0, \ldots, r+s-1$,

$$f_i(x) = \frac{x^i}{(x-\alpha)^r (x-\beta)^s}$$

so that $f_0, \ldots, f_{r+s-1}$ is a basis for E.

As for the second part, let $B = \{g_1, \ldots, g_r, h_1, \ldots, h_s\}$. Then B is linearly independent. In fact, suppose that, for every $x \in \mathbb{R} \setminus \{\alpha, \beta\}$,

$$\frac{\lambda_1}{x-\alpha} + \cdots + \frac{\lambda_r}{(x-\alpha)^r} + \frac{\mu_1}{x-\beta} + \cdots + \frac{\mu_s}{(x-\beta)^s} = 0.$$

Multiplying both sides by $(x-\alpha)^r (x-\beta)^s$ we obtain

$$\lambda_1 (x-\alpha)^{r-1}(x-\beta)^s + \lambda_2 (x-\alpha)^{r-2}(x-\beta)^s + \cdots + \lambda_r (x-\beta)^s +$$
$$\mu_1 (x-\alpha)^r (x-\beta)^{s-1} + \mu_2 (x-\alpha)^r (x-\beta)^{s-2} + \cdots + \mu_s (x-\alpha)^r = 0.$$

Taking the term $\lambda_r (x-\beta)^s$ over to the right hand side, what remains on the left is divisible by $x-\alpha$ and, since $\alpha \neq \beta$, we deduce that $\lambda_r = 0$; similarly we see that $\mu_s = 0$. Extracting a resulting factor $(x-\alpha)(x-\beta)$ we can repeat this argument to obtain $\lambda_{r-1} = 0 = \mu_{s-1}$. Continuing in this way, we obtain every $\lambda_i = 0$ and every $\mu_j = 0$ whence we see that B is linearly independent. Since B has $r+s$ elements and since the dimension of E is $r+s$ it follows that B is a basis for E.

4.22 If (*a*) holds then $\emptyset$ is the only basis for V. If (*b*) holds then $\{v\}$ is a basis for some non-zero $v \in V$. Since every $x \in V$ can be written uniquely in the form $x = \lambda v$ for some $\lambda \in F$ and since $F = \{0, 1\}$, it follows that $V = \{0, v\}$ whence $\{v\}$ is the only basis for V.

Conversely, suppose that V has precisely one basis and that $V \neq \{0\}$. Since for $x \neq 0$, $\{x\}$ is linearly independent it can be extended to form a basis (*the* basis of V). It follows that every non-zero element of V belongs to the basis. As V is of finite dimension, V is therefore finite. Suppose that $V = \{0, v_1, \ldots, v_n\}$ where $\{v_1, \ldots, v_n\}$ is the basis. Suppose further that $n \geqslant 2$ and consider the element $v_1 + \cdots + v_n$. Now $v_1 + \cdots + v_n \neq 0$ since $\{v_1, \ldots, v_n\}$ is linearly independent. Thus, for some i, we have $v_1 + \cdots + v_n = v_i$ whence

$$v_1 + \cdots + v_{i-1} + v_{i+1} + \cdots + v_n = 0.$$

This is impossible since every subset of $\{v_1, \ldots, v_n\}$ is linearly independent. This contradiction shows that we must have $n = 1$ and hence that $V = \{0, v_1\}$. It follows that (*b*) holds.

4.23 The result is trivial if $n = 1$, since f_1 is a non-zero element of Map(IR, IR). By way of induction, suppose that $\{f_1, \ldots, f_{n-1}\}$ is linearly independent whenever $r_1, \ldots, r_{n-1}$ are distinct. Consider $\{f_1, \ldots, f_n\}$ and suppose that $r_1, \ldots, r_n$ are distinct. If $\lambda_1 f_1 + \cdots + \lambda_n f_n = 0$ then

$$(\forall x \in \mathrm{IR}) \quad \lambda_1 \mathrm{e}^{r_1 x} + \ldots + \lambda_n \mathrm{e}^{r_n x} = 0.$$

Dividing by $\mathrm{e}^{r_n x}$ (which is non-zero) and differentiating, we obtain

$$\lambda_1(r_1 - r_n)\,\mathrm{e}^{(r_1 - r_n)x} + \ldots + \lambda_{n-1}(r_{n-1} - r_n)\,\mathrm{e}^{(r_{n-1} - r_n)x} = 0.$$

Since the $n-1$ real numbers $r_1 - r_n, \ldots, r_{n-1} - r_n$ are distinct, the induction hypothesis shows that $\lambda_1 = \cdots = \lambda_{n-1} = 0$. Consequently $\lambda_n f_n = 0$ and hence $\lambda_n = 0$ (since $\mathrm{e}^{r_n x} \neq 0$). Thus $\{f_1, \ldots, f_n\}$ is linearly independent. By induction, therefore, we have shown that if $r_1, \ldots, r_n$ are distinct then $\{f_1, \ldots, f_n\}$ is linearly independent. As for the converse, it suffices to note that if $r_1, \ldots, r_n$ are not distinct, say $r_i = r_j$ for $i \neq j$, then $f_i = f_j$ and $\{f_1, \ldots, f_n\}$ is dependent since

$$0f_1 + \cdots + 1f_i + \cdots + (-1)f_j + \cdots + 0f_n = 0.$$

Solutions to Chapter 5

5.1 (*a*), (*c*), (*f*) are linear; (*b*), (*d*), (*e*) are not. For example,

(*b*) $f(1,0,0)+f(-1,0,0)=(1,0,0)+(1,0,0)=(2,0,0)$ but $f[(1,0,0)+(-1,0,0)]=f(0,0,0)=(0,0,0)$;

(*d*) $f(0,0,0)=(-1,0,0)$ so $f[2(0,0,0)]\neq 2f(0,0,0)$;

(*e*) as in (*d*).

5.2 (*a*) We have that

$$\begin{aligned}f(0,0,1)&=f(1,1,2)-f(1,1,1)=(2,2,4)-(1,1,1)=(1,1,3),\\ f(0,1,1)&=f(1,2,3)-f(1,1,2)=(-1,-2,-3)-(2,2,4)\\ &=(-3,-4,-7),\end{aligned}$$

and hence

$$\begin{aligned}f(0,1,0)&=f(0,1,1)-f(0,0,1)=(-4,-5,-10),\\ f(1,0,0)&=f(1,1,1)-f(0,1,1)=(4,5,8).\end{aligned}$$

Consequently we have

$$\begin{aligned}f(a,b,c)&=af(1,0,0)+bf(0,1,0)+cf(0,0,1)\\ &=(4a-4b+c,5a-5b+c,8a-10b+3c).\end{aligned}$$

(*b*) It is not possible to find f in this case. The reason for this is that $\{(1,1,1),(1,1,2),(2,2,3)\}$ does not span $\mathbb{R}^3$. For example, $(1,0,0)$ is not a linear combination of these vectors so $f(1,0,0)$ cannot be determined using the given information.

(*c*) No such linear mapping exists. Indeed, if such a linear mapping f existed then since

$$f[(0,1,1)+(1,1,0)+(1,0,1)]=f(2,2,2)$$

we would have

$$f(0,1,1)+f(1,1,1)+f(1,0,1)=2f(1,1,1)$$

which is not the case.

5.3 (*a*), (*b*), (*c*) are linear. (*d*) is not linear; for example, we have $T(\lambda I)=\lambda B^2-\lambda^2 B$ whereas $\lambda T(I)=\lambda B^2-\lambda B$. As for (*e*), we note that $T(0)=7B^2$ so if T is linear we must have $B^2=0$. Conversely, if $B^2=0$ then

$$T(A)=AB+BA-2BA-3AB=-2AB-BA$$

which is linear. Hence $B^2=0$ is a necessary and sufficient condition for T to be linear.

5.4 We have

$$f(1,0)=(1,2,-1)=1(1,0,0)+2(0,1,0)-1(0,0,1),$$
$$f(0,1)=(2,-1,0)=2(1,0,0)-1(0,1,0)+0(0,0,1),$$

and consequently

$$A=\begin{bmatrix}1 & 2\\ 2 & -1\\ -1 & 0\end{bmatrix}.$$

Likewise, we have

$$g(1,0,0)=(2,0)=2(1,0)+0(0,1),$$
$$g(0,1,0)=(-1,2)=-1(1,0)+2(0,1),$$
$$g(0,0,1)=(0,-1)=0(1,0)-1(0,1),$$

and consequently

$$B=\begin{bmatrix}2 & -1 & 0\\ 0 & 2 & -1\end{bmatrix}.$$

To find C we note that

$$f(0,1)=(2,-1,0)=1(0,0,1)-3(0,1,1)+2(1,1,1),$$
$$f(1,1)=(3,1,-1)=-2(0,0,1)-2(0,1,1)+3(1,1,1),$$

so that

$$C=\begin{bmatrix}1 & -2\\ -3 & -2\\ 2 & 3\end{bmatrix}.$$

Finally, to find D we note that

$$g(0,0,1)=(0,-1)=-1(0,1)+0(1,1),$$

$$g(0,1,1) = (-1,1) = 2(0,1) + -1(1,1),$$
$$g(1,1,1) = (1,1) = 0(0,1) + 1(1,1),$$

and so

$$D = \begin{bmatrix} -1 & 2 & 0 \\ 0 & -1 & 1 \end{bmatrix}.$$

5.5 We have

$$f(0,1,0) = f(1,1,0) - f(1,0,0) = (4,1,4) - (2,3,-2) = (2,-2,6)$$
$$f(0,0,1) = f(1,1,1) - f(1,1,0) = (5,-1,7) - (4,1,4) = (1,-2,3)$$

Hence the matrix of f with respect to the standard basis of $\mathbb{R}^3$ is

$$\begin{bmatrix} 2 & 2 & 1 \\ 3 & -2 & -2 \\ -2 & 6 & 3 \end{bmatrix}.$$

5.6 We have

$$f(1,-1,0) = (2,-2,0) = 2(1,-1,0) + 0(1,0,-1) + 0(1,0,0),$$
$$f(1,0,-1) = (1,-2,-3) = 2(1,-1,0) + 3(1,0,-1) - 4(1,0,0),$$
$$f(1,0,0) = (2,-1,0) = 1(1,-1,0) + 0(1,0,-1) + 1(1,0,0),$$

and so the matrix is

$$\begin{bmatrix} 2 & 2 & 1 \\ 0 & 3 & 0 \\ 0 & -4 & 1 \end{bmatrix}.$$

5.7 To determine $p_{M,N}(x,y)$ in each of the given cases, we apply simple co-ordinate geometry.

(*a*) $p_{M,N}(x,y) = \left(\dfrac{y-nx}{m-n}, \dfrac{m(y-nx)}{m-n}\right).$

See Fig. S5.1.

(*b*) $p_{M,N}(x,y) = (x, mx).$
See Fig. S5.2.

(*c*) $p_{M,N}(x,y) = (0, y-nx).$
See Fig. S5.3.

Fig.S5.1

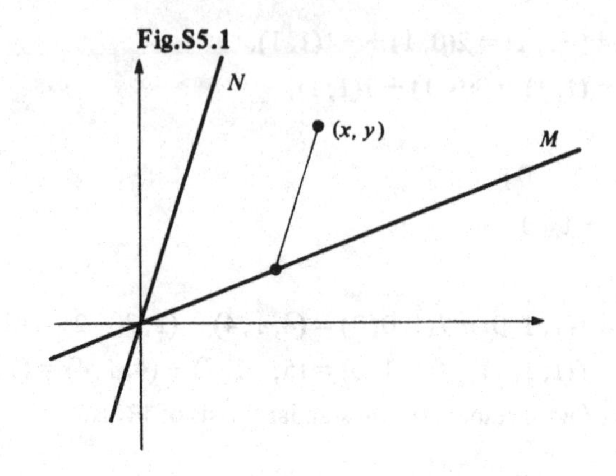

Fig.S5.2

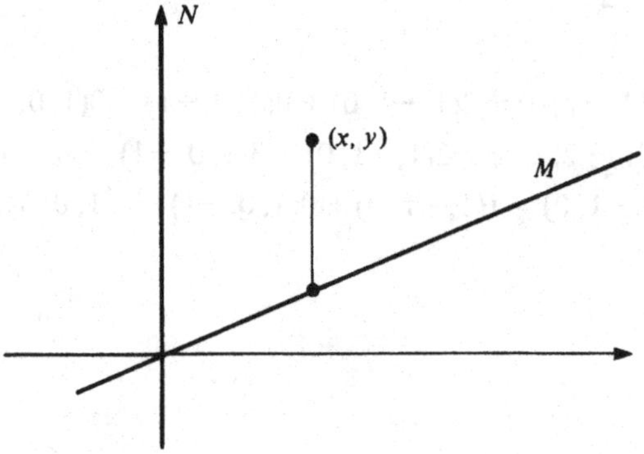

Fig.S5.3

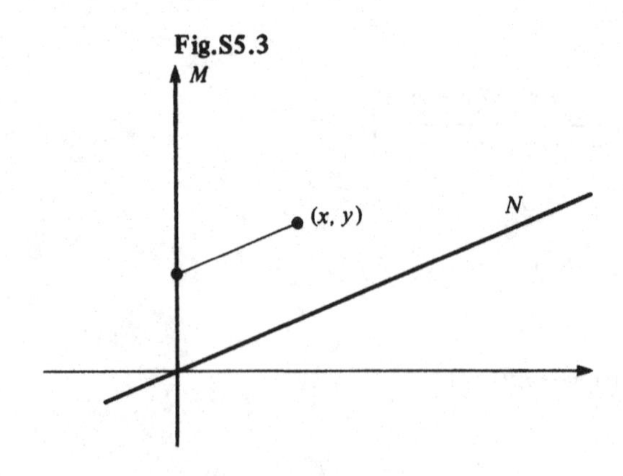

It is clear that in all three cases $p_{M,N}$ is linear. The corresponding matrices (relative to the standard basis of IR^2) are

$$\frac{1}{m-n}\begin{bmatrix}-n & 1\\ -mn & m\end{bmatrix},\quad \begin{bmatrix}1 & 0\\ m & 0\end{bmatrix},\quad \begin{bmatrix}0 & 0\\ -n & 1\end{bmatrix}.$$

It is readily seen that the characteristic polynomial of each of these is $X(X-1)$ and that this is also the minimum polynomial.

5.8 Suppose that f is injective and that $\{v_1, \ldots, v_n\}$ is linearly independent. Then from

$$0 = \sum_{i=1}^{n} \lambda_i f(v_i) = f\left(\sum_{i=1}^{n} \lambda_i v_i\right)$$

we deduce that $\sum_{i=1}^{n} \lambda_i v_i = 0$ whence each λ_i is zero and consequently $\{f(v_1), \ldots, f(v_n)\}$ is linearly independent.

Conversely, suppose that $\{v_1, \ldots, v_n\}$ linearly independent implies that $\{f(v_1), \ldots, f(v_n)\}$ is linearly independent. For any non-zero x we have that $\{x\}$ is linearly independent. Hence so also is $\{f(x)\}$, whence $f(x) \neq 0$. The property $x \neq 0 \Rightarrow f(x) \neq 0$ is equivalent to f being injective.

5.9 That f_y is linear is a routine verification of the axioms. Clearly we have $x \in \operatorname{Ker} f_y$ if and only if

$$x_2y_3 - x_3y_2 = x_3y_1 - x_1y_3 = x_1y_2 - x_2y_1 = 0.$$

Equally clearly we have $y \in \operatorname{Ker} f_y$ and so the subspace generated by $\{y\}$ is contained in the subspace $\operatorname{Ker} f_y$. To obtain the reverse inclusion, let $x \in \operatorname{Ker} f_y$. Since $y \neq 0$ we can suppose without loss of generality that $y_1 \neq 0$. If $y_2 = y_3 = 0$ then from the above equalities we have $x_2 = x_3 = 0$ whence x is a scalar multiple of y. If $y_2 \neq 0 = y_3$ then $x_3 = 0$ and $x_1/y_1 = x_2/y_2 = \lambda$, say, giving $x_1 = \lambda y_1$, $x_2 = \lambda y_2$ and again x is a scalar multiple of y. The other cases are argued similarly. The outcome is that x belongs to the subspace generated by $\{y\}$, and the reverse inclusion is established.

5.10 Since f is linear so also is ϑ; in fact

$$\begin{aligned}\vartheta(x,y) + \vartheta(x',y') &= (x, y-f(x)) + (x', y'-f(x'))\\ &= (x+x', y+y'-f(x+x')) = \vartheta(x+x', y+y');\\ \vartheta(\lambda x, \lambda y) &= (\lambda x, \lambda y - f(\lambda x))\\ &= \lambda(x, y-f(x)) = \lambda\vartheta(x,y).\end{aligned}$$

Clearly, we have $\vartheta(x,y) = (0,0)$ if and only if $x = 0$ and $y = f(x)$, which is the case if and only if $x = y = 0$. Thus we have $\text{Ker}\, \vartheta = \{(0,0)\}$ whence ϑ is injective. To see that ϑ is also surjective we observe that

$$\vartheta(x, y + f(x)) = (x, y + f(x) - f(x)) = (x, y).$$

Hence ϑ is an isomorphism.

5.11 (*a*) ⇒ (*b*): If $\text{Im}\, f = \text{Ker}\, f$ then for every $x \in V$ we have $f(x) \in \text{Im}\, f = \text{Ker}\, f$ so that $f[f(x)] = 0$ and hence $f \circ f = 0$. Moreover, $f \neq 0$ since the image of the zero map is $\{0\}$ and the kernel of the zero map is V. Now $n = \dim V = \dim \text{Im}\, f + \dim \text{Ker}\, f$ so from (*a*) we also deduce that $n = 2\, \text{rank}\, f$ whence n is even and $\text{rank}\, f = \frac{1}{2}n$.

(*b*) ⇒ (*a*): Suppose now that (*b*) holds. From $f \circ f = 0$ we have $f[f(x)] = 0$ for every $x \in V$ whence $\text{Im}\, f \subseteq \text{Ker}\, f$. Since $f \neq 0$, n is even and $\text{rank}\, f = \frac{1}{2}n$. We deduce from $n = \dim V = \dim \text{Im}\, f + \dim \text{Ker}\, f$ that $\dim \text{Im}\, f = \dim \text{Ker}\, f = \frac{1}{2}n$ whence we conclude that $\text{Im}\, f = \text{Ker}\, f$.

5.12 Using the fact that $\dim V_i = \dim \text{Im}\, f_i + \dim \text{Ker}\, f_i$ we have

$$\begin{aligned} -\dim V_1 &= -\dim \text{Im}\, f_1 - \dim \text{Ker}\, f_1 \\ \dim V_2 &= \dim \text{Im}\, f_2 + \dim \text{Ker}\, f_2 \\ -\dim V_3 &= -\dim \text{Im}\, f_3 - \dim \text{Ker}\, f_3 \\ &\vdots \\ (-1)^n \dim V_n &= (-1)^n \dim \text{Im}\, f_n + (-1)^n \dim \text{Ker}\, f_n \end{aligned}$$

Now it is given that $\text{Ker}\, f_1 = \{0\}$, $\text{Im}\, f_n = V_{n+1}$ and $\text{Im}\, f_i = \text{Ker}\, f_{i+1}$ so summing the above equalities, we obtain

$$\sum_{i=1}^{n} (-1)^i \dim V_i = (-1)^n \dim V_{n+1}$$

and hence

$$\sum_{i=1}^{n+1} (-1)^i \dim V_i = 0.$$

5.13 If $x \neq 0$ is such that $f^{p-1}(x) \neq 0$ then $x, f(x), \ldots, f^{p-1}(x)$ are all non-zero for $f^k(x) = 0$ with $k \leqslant p-1$ implies $f^{p-1}(x) = f^{p-1-k}[f^k(x)] = 0$, a contradiction. To show that the set $\{x, f(x), \ldots, f^{p-1}(x)\}$ is linearly independent, suppose that

$$\lambda_0 x + \lambda_1 f(x) + \cdots + \lambda_{p-1} f^{p-1}(x) = 0.$$

Applying f^{p-1} to this we obtain $\lambda_0 f^{p-1}(x) = 0$ whence $\lambda_0 = 0$ since

$f^{p-1}(x) \neq 0$. Thus we have

$$\lambda_1 f(x) + \cdots + \lambda_{p-1} f^{p-1}(x) = 0.$$

Applying f^{p-2} to this we obtain similarly $\lambda_1 = 0$. Continuing in this way, we see that every λ_i is zero and that consequently the set is linearly independent.

If f is nilpotent of index $n = \dim V$ then clearly $\{x, f(x), \ldots, f^{n-1}(x)\}$ is a basis of V. The images of these basis vectors under f are given by

$$\begin{aligned}
f(x) &= 0x + 1f(x) + 0f^2(x) + \cdots + 0f^{n-1}(x)\\
f^2(x) &= 0x + 0f(x) + 1f^2(x) + \cdots + 0f^{n-1}(x)\\
&\vdots\\
f^{n-1}(x) &= 0x + 0f(x) + 0f^2(x) + \cdots + 1f^{n-1}(x)\\
f^n(x) &= 0x + 0f(x) + 0f^2(x) + \cdots + 0f^{n-1}(x)
\end{aligned}$$

from which we see that the matrix of f relative to this ordered basis is

$$M = \begin{bmatrix} 0 & 0 & 0 & \ldots & 0 & 0\\ 1 & 0 & 0 & \ldots & 0 & 0\\ 0 & 1 & 0 & \ldots & 0 & 0\\ \vdots & \vdots & \vdots & & \vdots & \vdots\\ 0 & 0 & 0 & \ldots & 1 & 0 \end{bmatrix}.$$

Conversely, suppose that there is an ordered basis with respect to which the matrix M of f is of the above form. It is readily verified that $M^n = 0$ and $M^{n-1} \neq 0$. Consequently f is nilpotent of index n.

For the last part it suffices to observe that if A is an $n \times n$ matrix which is similar to the above matrix M then there is an invertible matrix P such that $P^{-1}AP = M$. Since then $A = PMP^{-1}$ we have (by induction) $A^r = PM^rP^{-1}$ for every positive integer r. In particular, $A^{n-1} = PM^{n-1}P^{-1} \neq 0$ and $A^n = PM^nP^{-1} = POP^{-1} = 0$.

Conversely, if M is nilpotent of index n and if $f : V \to V$ is such that a matrix of f is M then f has a matrix of the required form which is similar to M.

5.14 Let $f : \mathbb{C}^2 \to \mathbb{C}^2$ be the linear transformation that is represented, relative to the canonical basis of $\mathbb{C}^2$, by the matrix

$$\begin{bmatrix} \cos\vartheta & -\sin\vartheta\\ \sin\vartheta & \cos\vartheta \end{bmatrix}.$$

Then for all $(x, y) \in \mathbb{C}^2$ we have

$$f(x, y) = (x \cos \vartheta - y \sin \vartheta, x \sin \vartheta + y \cos \vartheta).$$

We wish to find $\alpha_1, \alpha_2 \in \mathbb{C}^2$ such that $f(\alpha_1) = e^{i\vartheta}\alpha_1$ and $f(\alpha_2) = e^{-i\vartheta}\alpha_2$ with $\{\alpha_1, \alpha_2\}$ a basis of $\mathbb{C}^2$. It is readily seen that $\alpha_1 = (i, 1)$ and $\alpha_2 = (1, i)$ satisfy these properties. The matrix of f relative to the basis $\{\alpha_1, \alpha_2\}$ is then

$$\begin{bmatrix} e^{i\vartheta} & 0 \\ 0 & e^{-i\vartheta} \end{bmatrix}$$

which is therefore similar to the first matrix.

5.15 For every linear transformation $f : V \to V$ we have $f(0) = 0$. The mapping described in (*a*) and (*c*) are not linear, therefore, since in each case we have $f(0, 0, 0) \neq (0, 0, 0)$. That described by (*b*) is linear. In fact $\mathbb{Z}_2 = \{0, 1\}$ and $a^2 = a$ for every $a \in \mathbb{Z}_2$, so that $f(a, b, c) = (a, b, c)$, i.e. f is the identity map.

5.16 f can be described by $f(a, b) = (a + b, a)$ and is linear.

5.17 D is linear since

$$(\forall p, q \in \mathbb{R}[X])(\forall a, b \in \mathbb{R}) \quad \mathrm{D}(ap + bq) = a\mathrm{D}p + b\mathrm{D}q.$$

T is also linear since

$$T(ap + bq) = X(ap + bq) = aXp + bXq = aT(p) + bT(q).$$

It is readily seen that

$$(\mathrm{D} + T)(a_0 + \cdots + a_nX^n) =$$
$$a_1 + (a_0 + 2a_2)X + \cdots + (a_{n-2} + na_n)X^{n-1} + a_{n-1}X^n + a_nX^{n+1};$$
$$(\mathrm{D} \circ T)(a_0 + \cdots + a_nX^n) = a_0 + 2a_1X + \cdots + (n + 1)a_nX^n;$$
$$(T \circ \mathrm{D})(a_0 + \cdots + a_nX^n) = a_1X + 2a_2X^2 + \cdots + na_nX^n.$$

Since $(\mathrm{D} \circ T)(p) = p + X\mathrm{D}(p) = p + (T \circ \mathrm{D})(p)$ we see that $\mathrm{D} \circ T - T \circ \mathrm{D} =$ It follows that, writing $T\mathrm{D}$ for $T \circ D$ etc.,

$$(T\mathrm{D})^2 = T\mathrm{D}T\mathrm{D} = T(I + T\mathrm{D})\mathrm{D} = T\mathrm{D} + T^2\mathrm{D}^2.$$

Define S by

$$S(a_0 + \cdots + a_nX^n) = a_0X + \frac{a_1}{2}X^2 + \cdots + \frac{a_n}{n+1}X^{n+1}.$$

Then it is readily seen that S is linear and that $\mathrm{D} \circ S = I$. Now D is not an isomorphism since it fails to be injective (for example $\mathrm{D}(1) = \mathrm{D}(2) = 0$); and S is not an isomorphism since it fails to be surjective (for example, there is no p with $S(p) = 1$).

5.18 (*a*) When $F = \mathbb{R}$ we have

$$(a, b, c) \in \operatorname{Ker} f \Leftrightarrow 0 = a + b = a + c = b + c$$
$$\Leftrightarrow a = b = c = 0.$$

Thus $\operatorname{Ker} f = 0$ and so $\dim \operatorname{Ker} f = 0$. Also, f is surjective since

$$f(\tfrac{1}{2}(a + b - c), \tfrac{1}{2}(a - b + c), \tfrac{1}{2}(-a + b + c)) = (a, b, c).$$

Thus $\operatorname{Im} f = \mathbb{R}^3$ and $\dim \operatorname{Im} f = 3$.

(*b*) When $F = \mathbb{Z}_2$ we have

$$(a, b, c) \in \operatorname{Ker} f \Leftrightarrow 0 = a + b = a + c = b + c$$
$$\Leftrightarrow a = b = c.$$

Thus $\operatorname{Ker} f = \{(0, 0, 0), (1, 1, 1)\} = \{(a, a, a) \mid a \in \mathbb{Z}_2\}$, and so in this case $\dim \operatorname{Ker} f = 1$. It follows that $\dim \operatorname{Im} f = 3 - 1 = 2$. Now $f(1, 0, 0) = (1, 1, 0)$ and $f(0, 1, 0) = (1, 0, 1)$ are linearly independent elements of $\operatorname{Im} f$ so they form a basis, whence

$$\operatorname{Im} f = \{(0, 0, 0), (1, 1, 0), (0, 1, 0), (0, 1, 1)\}.$$

5.19 We have $f(a, b, c) = (b, -a, c)$ so

$$f(1, 0, 0) = (0, -1, 0) = 0(1, 0, 0) - 1(0, 1, 0) + 0(0, 0, 1),$$
$$f(0, 1, 0) = (1, 0, 0) = 1(1, 0, 0) + 0(0, 1, 0) + 0(0, 0, 1),$$
$$f(0, 0, 1) = (0, 0, 1) = 0(1, 0, 0) + 0(0, 1, 0) + 1(0, 0, 1),$$

and consequently

$$A = \begin{bmatrix} 0 & 1 & 0 \\ -1 & 0 & 0 \\ 0 & 0 & 1 \end{bmatrix}.$$

Similarly, we have

$$f(1, 1, 0) = (1, -1, 0) = 0(1, 1, 0) - 1(0, 1, 1) + 1(1, 0, 1),$$
$$f(0, 1, 1) = (1, 0, 1) = 0(1, 1, 0) + 0(0, 1, 1) + 1(1, 0, 1),$$
$$f(1, 0, 1) = (0, -1, 1) = -1(1, 1, 0) + 0(0, 1, 1) + 1(1, 0, 1),$$

and consequently

$$B = \begin{bmatrix} 0 & 0 & -1 \\ -1 & 0 & 0 \\ 1 & 1 & 1 \end{bmatrix}.$$

The matrix X represents the identity map relative to a change of reference from the first basis to the second. Since

$$(1,0,0) = \tfrac{1}{2}(1,1,0) - \tfrac{1}{2}(0,1,1) + \tfrac{1}{2}(1,0,1),$$
$$(0,1,0) = \tfrac{1}{2}(1,1,0) + \tfrac{1}{2}(0,1,1) - \tfrac{1}{2}(1,0,1),$$
$$(0,0,1) = -\tfrac{1}{2}(1,1,0) + \tfrac{1}{2}(0,1,1) + \tfrac{1}{2}(1,0,1),$$

we see that

$$X = \begin{bmatrix} \frac{1}{2} & \frac{1}{2} & -\frac{1}{2} \\ -\frac{1}{2} & \frac{1}{2} & \frac{1}{2} \\ \frac{1}{2} & -\frac{1}{2} & \frac{1}{2} \end{bmatrix}.$$

5.20 The matrix of f with respect to $\{u_1, u_2\}$ is $\begin{bmatrix} 1 & 1 \\ -1 & 0 \end{bmatrix}$. The matrix representin the change of basis is $\begin{bmatrix} 3 & 1 \\ -1 & 1 \end{bmatrix}$, the inverse of which is $\begin{bmatrix} \frac{1}{4} & -\frac{1}{4} \\ \frac{1}{4} & \frac{3}{4} \end{bmatrix}$. Hence th matrix of f relative to $\{w_1, w_2\}$ is

$$\begin{bmatrix} \frac{1}{4} & -\frac{1}{4} \\ \frac{1}{4} & \frac{3}{4} \end{bmatrix} \begin{bmatrix} 1 & 1 \\ -1 & 0 \end{bmatrix} \begin{bmatrix} 3 & 1 \\ -1 & 1 \end{bmatrix} = \begin{bmatrix} \frac{5}{4} & \frac{3}{4} \\ -\frac{7}{4} & -\frac{1}{4} \end{bmatrix}.$$

5.21 The respective matrices are

$$A = \begin{bmatrix} 0 & 1 & 0 & 0 & \dots & 0 \\ 0 & 0 & 2 & 0 & \dots & 0 \\ 0 & 0 & 0 & 3 & \dots & 0 \\ \vdots & \vdots & \vdots & \vdots & & \vdots \\ 0 & 0 & 0 & 0 & \dots & n-1 \\ 0 & 0 & 0 & 0 & \dots & 0 \end{bmatrix},$$

$$B = \begin{bmatrix} 0 & 0 & 0 & \dots & 0 & 0 \\ n-1 & 0 & 0 & \dots & 0 & 0 \\ 0 & n-2 & 0 & \dots & 0 & 0 \\ 0 & 0 & n-3 & \dots & 0 & 0 \\ \vdots & \vdots & \vdots & & \vdots & \vdots \\ 0 & 0 & 0 & \dots & 1 & 0 \end{bmatrix},$$

$$C = \begin{bmatrix} 0 & 1 & -2 & -3 & \dots & -n+1 \\ 0 & 0 & 2 & 0 & \dots & 0 \\ 0 & 0 & 0 & 3 & \dots & 0 \\ \vdots & \vdots & \vdots & \vdots & & \vdots \\ 0 & 0 & 0 & 0 & \dots & n-1 \\ 0 & 0 & 0 & 0 & \dots & 0 \end{bmatrix}.$$

5.22 For every $z \in \mathbb{C}$ we have $z = a.1 + b.\mathrm{i}$ for some $a, b \in \mathrm{IR}$; and if $\lambda.1 + \mu.\mathrm{i} = 0$ then $\lambda = \mu = 0$. Thus $\{1, \mathrm{i}\}$ is a basis for $\mathbb{C}$ as a vector space over IR. Since

$$\begin{aligned} f(\alpha(x + \mathrm{i}y) + \beta(a + \mathrm{i}b)) &= (\alpha x + \beta a) - \mathrm{i}(\alpha y + \beta b) \\ &= \alpha f(x + \mathrm{i}y) + \beta f(a + \mathrm{i}b) \end{aligned}$$

it follows that f is linear. The matrix of f relative to the basis $\{1, \mathrm{i}\}$ is

$$\begin{bmatrix} 1 & 0 \\ 0 & -1 \end{bmatrix}.$$

Regarding $\mathbb{C}$ as a vector space over $\mathbb{C}$, we see that it is of dimension 1 with basis $\{1\}$. In this situation f is not linear. For example, take $\alpha = 1 - \mathrm{i}$ and $z = 1 + \mathrm{i}$; we have

$$f(\alpha z) = 2 \quad \text{but} \quad \alpha f(z) = -2\mathrm{i}.$$

5.23 Consider the following row reduction, in which we have been careful not to divide by either 2 or 3 (since these are 0 in $\mathbb{Z}_2$ and $\mathbb{Z}_3$ respectively):

$$\begin{bmatrix} 3 & -1 & 1 \\ -1 & 5 & -1 \\ 1 & -1 & 3 \end{bmatrix} \rightarrow \begin{bmatrix} 1 & -1 & 3 \\ -1 & 5 & -1 \\ 3 & -1 & 1 \end{bmatrix}$$

$$\rightarrow \begin{bmatrix} 1 & -1 & 3 \\ 0 & 4 & 2 \\ 0 & 2 & -8 \end{bmatrix} \rightarrow \begin{bmatrix} 1 & -1 & 3 \\ 0 & 2 & -8 \\ 0 & 4 & 2 \end{bmatrix}$$

$$\rightarrow \begin{bmatrix} 1 & -1 & 3 \\ 0 & 2 & -8 \\ 0 & 0 & 18 \end{bmatrix}.$$

(*a*) When $F = \mathrm{IR}$ the rank of this final matrix is 3, and so $\mathrm{Im}\, f$ is of dimension 3 and $\mathrm{Ker}\, f$ is of dimension 0.

(*b*) When $F = \mathbb{Z}_2$ we have that 2, 18, −8 are all zero, in which case the rank of the matrix reduces to 1. So in this case we have $\operatorname{Im} f$ is of dimension 1 and $\operatorname{Ker} f$ is of dimension 2.

(*c*) When $F = \mathbb{Z}_3$ we have that 18 is zero, in which case the rank of the matrix reduces to 2. So in this case we have $\operatorname{Im} f$ is of dimension 2 and $\operatorname{Ker} f$ is of dimension 1.

5.24 If $f : \mathrm{IR}^2 \to \mathrm{IR}^2$ is given by $f(a, b) = (b, 0)$ then we have

$$\operatorname{Im} f = \operatorname{Ker} f = \{(a, 0) \mid a \in \mathrm{IR}\}.$$

If $g : \mathrm{IR}^3 \to \mathrm{IR}^3$ is given by $g(a, b, c) = (c, 0, 0)$ then

$$\{(a, 0, 0) \mid a \in \mathrm{IR}\} = \operatorname{Im} g \subset \operatorname{Ker} g = \{(a, b, 0) \mid a, b \in \mathrm{IR}\}.$$

If $h : \mathrm{IR}^3 \to \mathrm{IR}^3$ is given by $h(a, b, c) = (b, c, 0)$ then

$$\{(a, 0, 0) \mid a \in \mathrm{IR}\} = \operatorname{Ker} h \subset \operatorname{Im} h = \{(a, b, 0) \mid a, b \in \mathrm{IR}\}.$$

5.25 Consider the following row reduction in which we have been careful not to divide by any integer n:

$$\begin{bmatrix} 2 & 2 & 1 \\ 1 & 3 & 1 \\ 1 & 2 & 2 \end{bmatrix} \to \begin{bmatrix} 1 & 3 & 1 \\ 2 & 2 & 1 \\ 1 & 2 & 2 \end{bmatrix}$$

$$\to \begin{bmatrix} 1 & 3 & 2 \\ 0 & -4 & -1 \\ 0 & -1 & 1 \end{bmatrix} \to \begin{bmatrix} 1 & 3 & 2 \\ 0 & -1 & 1 \\ 0 & -4 & -1 \end{bmatrix}$$

$$\to \begin{bmatrix} 1 & 3 & 2 \\ 0 & -1 & 1 \\ 0 & 0 & -5 \end{bmatrix}$$

It is clear from this final matrix that f is an isomorphism if and only if the matrix is of full rank 3; and this is the case if and only if $p \neq 5$.

Solutions to Chapter 6

6.1 Since V is a real inner product space we have $\langle x \mid y\rangle = \langle y \mid x\rangle$. Using the fact that $\|x\|^2 = \langle x \mid x\rangle$ we thus have

$$\left\langle y - \frac{\langle x \mid y\rangle}{\|x\|^2} x \,\middle|\, x\right\rangle = \langle y \mid x\rangle - \frac{\langle x \mid y\rangle}{\|x\|^2} \langle x \mid x\rangle = 0.$$

The second part of the question consists of finding a scalar λ such that $\langle y - \lambda x \mid x\rangle = 0$. Clearly, λ is unique and by the above we see that it is given by $\lambda = \langle x \mid y\rangle / \|x\|^2$.

We have

$$|\cos \vartheta| = \frac{\|\lambda x\|}{\|y\|} = \frac{|\lambda|\,\|x\|}{\|y\|} = \frac{|\langle x \mid y\rangle|}{\|x\|\,\|y\|}.$$

Since $\cos \vartheta \leqslant 0 \Leftrightarrow \lambda \leqslant 0$ it follows that $\cos \vartheta = \langle x \mid y\rangle / \|x\|\,\|y\|$.

6.2 Since V is a real inner product space we have

$$\begin{aligned}\|x+y\|^2 = \langle x+y \mid x+y\rangle &= \langle x \mid x\rangle + \langle y \mid x\rangle + \langle x \mid y\rangle + \langle y \mid y\rangle \\ &= \|x\|^2 + \|y\|^2 + 2\langle x \mid y\rangle.\end{aligned}$$

In $\mathbb{R}^2$ this is, by the previous question, the cosine law

$$\|x+y\|^2 = \|x\|^2 + \|y\|^2 + 2\|x\|\,\|y\| \cos \vartheta.$$

If now V is a complex inner product space then we have

$$\begin{aligned}\|x+y\|^2 - \mathrm{i}\|\mathrm{i}x+y\|^2 &= \|x\|^2 + \|y\|^2 + \langle x \mid y\rangle + \langle y \mid x\rangle \\ &\quad - \mathrm{i}\,[\|\mathrm{i}x\|^2 + \|y\|^2 + \langle \mathrm{i}x \mid y\rangle + \langle y \mid \mathrm{i}x\rangle] \\ &= \|x\|^2 + \|y\|^2 - \mathrm{i}(\|x\|^2 + \|y\|^2) \\ &\quad + \langle x \mid y\rangle + \langle y \mid x\rangle - \mathrm{i}^2\langle x \mid y\rangle + \mathrm{i}^2\langle y \mid x\rangle \\ &= \|x\|^2 + \|y\|^2 - \mathrm{i}(\|x\|^2 + \|y\|^2) + 2\langle x \mid y\rangle.\end{aligned}$$

6.3 We have

$$\|x+y\|^2+\|x-y\|^2=\|x\|^2+\|y\|^2+\langle x\mid y\rangle+\langle y\mid x\rangle$$
$$+\|x\|^2+\|-y\|^2+\langle x\mid -y\rangle+\langle -y\mid x\rangle$$
$$=2\|x\|^2+2\|y\|^2.$$

In $\mathbb{R}^2$ this result says that the sum of the squares of the lengths of th diagonals of a parallelogram is the sum of the squares of the lengths of it sides.

6.4 If $\|x\|=\|y\|$ then

$$\langle x+y\mid x-y\rangle=\langle x\mid x\rangle+\langle y\mid x\rangle+\langle x\mid -y\rangle+\langle y\mid -y\rangle$$
$$=\|x\|^2-\|y\|^2=0.$$

In $\mathbb{R}^2$ this result says that the diagonals of a rhombus are mutually perpe dicular.

6.5 The respective inequalities are

(*a*) $|\sum_{k=1}^n \alpha_k\bar{\beta}_k| \leqslant (\sum_{k=1}^n |\alpha_k|^2)^{1/2}(\sum_{k=1}^n |\beta_k|^2)^{1/2}$;

(*b*) $|\int_0^1 f(x)g(x)\,dx| \leqslant (\int_0^1 |f(x)|^2\,dx)^{1/2}(\int_0^1 |g(x)|^2\,dx)^{1/2}$.

6.6 We have that

$$\operatorname{tr}(A+B)=\sum_{i=1}^n (a_{ii}+b_{ii})=\sum_{i=1}^n a_{ii}+\sum_{i=1}^n b_{ii}=\operatorname{tr}(A)+\operatorname{tr}(B).$$

Also,

$$\operatorname{tr}(AB)=\sum_{i=1}^n [AB]_{ii}=\sum_{i=1}^n\sum_{j=1}^n a_{ij}b_{ji}$$

which, on interchanging i and j is the same as $\operatorname{tr}(BA)$.

That $\langle A\mid B\rangle=\operatorname{tr}(AB^*)$ defines an inner product follows from a caref verification of the axioms:

$$\langle A\mid B+C\rangle=\operatorname{tr}[A(B+C)^*]=\operatorname{tr}(AB^*+AC^*)$$
$$=\operatorname{tr}(AB^*)+\operatorname{tr}(AC^*)=\langle A\mid B\rangle+\langle A\mid C\rangle;$$
$$\langle A+B\mid C\rangle=\operatorname{tr}[(A+B)C^*]=\operatorname{tr}(AC^*+BC^*)$$
$$=\operatorname{tr}(AC^*)+\operatorname{tr}(BC^*)=\langle A\mid C\rangle+\langle B\mid C\rangle;$$
$$\langle\lambda A\mid B\rangle=\operatorname{tr}(\lambda AB^*)=\lambda\operatorname{tr}(AB^*)=\lambda\langle A\mid B\rangle;$$
$$\langle A\mid\lambda B\rangle=\operatorname{tr}[A(\lambda B)^*]=\operatorname{tr}[A\bar{\lambda}B^*]=\bar{\lambda}\langle A\mid B\rangle;$$
$$\overline{\langle B\mid A\rangle}=\overline{\operatorname{tr}(BA^*)}=\operatorname{tr}(AB^*)=\langle A\mid B\rangle;$$
$$\langle A\mid A\rangle=\operatorname{tr}(AA^*)=\sum_{i,j} a_{ij}\bar{a}_{ij}=\sum_{i,j}|a_{ij}|^2\geqslant 0$$
$$\text{with } \langle A\mid A\rangle=0\Leftrightarrow A=0.$$

The Cauchy-Schwarz inequality in this complex inner product space is $|\text{tr}\,(AB^*)| \leqslant |\text{tr}\,(AA^*)|^{1/2}\,|\text{tr}\,(BB^*)|^{1/2}$.

It is clear that $\{E_{pq}\}$ is a basis for V since every $A = [a_{ij}] \in V$ can be written uniquely in the form

$$A = \sum_{p=1}^{n} \sum_{q=1}^{n} a_{pq} E_{pq}.$$

To see that this is an orthonormal basis, we observe that

$$\begin{aligned} \langle E_{pq} \mid E_{rs} \rangle &= \text{tr}\,(E_{pq} E_{rs}^*) = \text{tr}\,(E_{pq} E_{sr}) \\ &= \begin{cases} 1 & \text{if } r = p, s = q; \\ 0 & \text{otherwise.} \end{cases} \end{aligned}$$

6.7 Let $x_1 = (1, 1, 0, 1)$, $x_2 = (1, -2, 0, 0)$, $x_3 = (1, 0, -1, 2)$. Then $\{x_1, x_2, x_3\}$ is linearly independent. Now $\|x_1\|^2 = 3$ so define $y_1 = (1/\sqrt{3})(1, 1, 0, 1)$. Then

$$x_2 - \langle x_2 \mid y_1 \rangle y_1 = \tfrac{1}{3}(4, -5, 0, 1)$$

so define $y_2 = (1/\sqrt{42})(4, -5, 0, 1)$. Then

$$x_3 - \langle x_3 \mid y_2 \rangle y_2 - \langle x_3 \mid y_1 \rangle y_1 = \tfrac{1}{7}(-4, -2, -1, 6)$$

so define $y_3 = (1/\sqrt{57})(-4, -2, -1, 6)$. By the Gram-Schmidt orthonormalisation process, $\{y_1, y_2, y_3\}$ is then an orthonormal basis for the subspace generated by $\{x_1, x_2, x_3\}$.

6.8 Let $f_1 : t \to 1$ and $f_2 : t \to t$. Then

$$\langle f_1 \mid f_1 \rangle = \int_0^1 [f_1(t)]^2\, \mathrm{d}t = \int_0^1 \mathrm{d}t = 1$$

so we can take $g_1 = f_1$ as the first vector in the Gram-Schmidt orthonormalisation process. Now

$$\langle f_2 \mid g_1 \rangle = \int_0^1 t\, \mathrm{d}t = \tfrac{1}{2}$$

and so $f_2 - \langle f_2 \mid g_1 \rangle g_1$ is the mapping $h : t \to t - \tfrac{1}{2}$. Now

$$\langle h \mid h \rangle = \int_0^1 [h(t)]^2\, \mathrm{d}t = \int_0^1 (t^2 - t + \tfrac{1}{4})\, \mathrm{d}t = \tfrac{1}{12}.$$

Thus we can define g_2 to be the mapping $t \to 2\sqrt{3}(t - \tfrac{1}{2})$. Then $\{g_1, g_2\}$ is an orthonormal basis.

6.9 It is clear that $X \subseteq Y \Rightarrow Y^\perp \subseteq X^\perp$. It follows that $V^\perp \subseteq (V \cap W)^\perp$ and $W^\perp \subseteq (V \cap W)^\perp$ whence $V^\perp + W^\perp \subseteq (V \cap W)^\perp$. Also, $(V+W)^\perp \subseteq V^\perp$ and $(V+W)^\perp \subseteq W^\perp$ whence $(V+W)^\perp \subseteq V^\perp \cap W^\perp$. Writing $V^\perp$, $W^\perp$ for V, W in the second of these observations, we obtain $(V^\perp + W^\perp)^\perp \subseteq V^{\perp\perp} \cap W^{\perp\perp} = V \cap W$ whence $(V \cap W)^\perp \subseteq (V^\perp + W^\perp)^{\perp\perp} = V^\perp + W^\perp$. Comparing this with the first observation, we see that $(V \cap W)^\perp = V^\perp + W^\perp$. Again writing $V^\perp$, $W^\perp$ for V, W and taking the $^\perp$ of each side, we obtain $V^\perp \cap W^\perp = (V+W)^\perp$.

6.10 That $\langle p \mid q \rangle = \int_0^1 p(X)q(X)\,\mathrm{d}X$ defines an inner product on $\mathbb{R}_3[X]$ is routine.

If $k \in K$ then

$$\langle k \mid q \rangle = k \int_0^1 (q_0 + q_1X + q_2X^2)\,\mathrm{d}X = k(q_0 + \tfrac{1}{2}q_1 + \tfrac{1}{3}q_2)$$

so q is orthogonal to every $k \in K$ if and only if $q_0 + \frac{1}{2}q_1 + \frac{1}{3}q_2 = 0$.

Since K is of dimension 1 we have $\dim K^\perp = \dim \mathbb{R}_3[X] - \dim K = 3 - 1 = 2$. A basis for $K^\perp$ is, for example, $\{p, q\}$ where $p(X) = 1 - 2X$, $q(X) = 1 - 3X^2$. We have $\langle p \mid p \rangle = \int_0^1 (1-2X)^2\,\mathrm{d}X = \frac{1}{3}$ so let $y_1(X) = \sqrt{3}(1-2X)$. Then $h = q - \langle q \mid y_1 \rangle y_1 = -\frac{1}{2} + 3X - 3X^2$ is such that $\|h\|^2 = \frac{1}{20}$ so we can take $y_2 = 2\sqrt{5}(-\frac{1}{2} + 3X - 3X^2)$ to complete an orthonormal basis for $K^\perp$.

6.11 Using the fact that $\operatorname{tr}(XY) = \operatorname{tr}(YX)$ we have

$$\begin{aligned}\langle f_M(A) \mid B \rangle = \langle MA \mid B \rangle &= \operatorname{tr}(MAB^*) = \operatorname{tr}(AB^*M) \\ &= \operatorname{tr}[A(M^*B)^*] = \langle A \mid M^*B \rangle = \langle A \mid f_{M^*}(B) \rangle.\end{aligned}$$

It follows from the uniqueness of adjoints that $f_M^* = f_{M^*}$.

6.12 We have that

$$\begin{aligned}\langle f_p(q) \mid r \rangle = \langle pq \mid r \rangle &= \int_0^1 (pq)(X)r(X)\,\mathrm{d}X \\ &= \int_0^1 p(X)q(X)r(X)\,\mathrm{d}X \\ &= \int_0^1 q(X)(pr)(X)\,\mathrm{d}X = \langle q \mid pr \rangle = \langle q \mid f_p(r) \rangle.\end{aligned}$$

It follows from the uniqueness of adjoints that $f_p^* = f_p$, so that f_p is self-adjoint.

6.13 If $y \in \operatorname{Im} f^*$, say $y = f^*(z)$, then

$$(\forall x \in \operatorname{Ker} f) \quad \langle x \mid y \rangle = \langle x \mid f^*(z) \rangle = \langle f(x) \mid z \rangle = \langle 0 \mid z \rangle = 0$$

and so $y \in (\operatorname{Ker} f)^\perp$. Thus $\operatorname{Im} f^* \subseteq (\operatorname{Ker} f)^\perp$.

If $y \in \operatorname{Ker} f^*$ then

$$(\forall x \in V) \quad \langle f(x) \mid y \rangle = \langle x \mid f^*(y) \rangle = \langle x \mid 0 \rangle = 0$$

and so $y \in (\operatorname{Im} f)^\perp$. Thus $\operatorname{Ker} f^* \subseteq (\operatorname{Im} f)^\perp$.

From the first inclusion we obtain

$$\dim \operatorname{Im} f^* \leqslant \dim (\operatorname{Ker} f)^\perp = \dim V - \dim \operatorname{Ker} f = \dim \operatorname{Im} f$$

so that

$$\dim \operatorname{Ker} f^* = \dim V - \dim \operatorname{Im} f^* \geqslant \dim V - \dim \operatorname{Im} f = \dim \operatorname{Ker} f. \tag{1}$$

From the second inclusion we obtain

$$\dim \operatorname{Ker} f^* \leqslant \dim (\operatorname{Im} f)^\perp = \dim V - \dim \operatorname{Im} f = \dim \operatorname{Ker} f. \tag{2}$$

It follows from (1) and (2) that $\dim \operatorname{Ker} f^* = \dim \operatorname{Ker} f = \dim (\operatorname{Im} f)^\perp$ and so, from the second inclusion, we deduce that $\operatorname{Ker} f^* = (\operatorname{Im} f)^\perp$. Likewise we can show that $\operatorname{Im} f^* = (\operatorname{Ker} f)^\perp$.

6.14 That $f_{x,y}$ is linear is routine. As for (*a*) we have

$$\begin{aligned}(f_{x,y} \circ f_{y,z})(t) &= f_{x,y}(\langle t \mid z \rangle y) \\ &= \langle \langle t \mid z \rangle y \mid y \rangle x \\ &= \langle t \mid z \rangle \langle y \mid y \rangle x \\ &= \|y\|^2 \langle t \mid z \rangle x \\ &= \|y\|^2 f_{x,z}(t).\end{aligned}$$

As for (*b*) we have

$$\begin{aligned}\langle f_{x,y}(z) \mid t \rangle &= \langle \langle z \mid y \rangle x \mid t \rangle \\ &= \langle z \mid y \rangle \langle x \mid t \rangle \\ &= \langle z \mid y \overline{\langle x \mid t \rangle} \rangle \\ &= \langle z \mid \langle t \mid x \rangle y \rangle \\ &= \langle z \mid f_{y,x}(t) \rangle.\end{aligned}$$

To establish (*c*) we recall that $f_{x,y}$ is normal if and only if it commutes with its adjoint. Using (*a*) and (*b*) we see that this is the case if and only if, for all $z \in V$,

$$\|y\|^2 \langle z \mid x \rangle x = \|x\|^2 \langle z \mid y \rangle y.$$

Taking $z = x$ in this we obtain

$$\|y\|^2\|x\|^2x = \|x\|^2\langle x \mid y\rangle y$$

which gives $x = \lambda y$ where $\lambda = \langle x \mid y\rangle/\|y\|^2 \in \mathbb{C}$, so that the condition is necessary. Conversely, if $x = \lambda y$ for some $\lambda \in \mathbb{C}$ then

$$\begin{aligned}\|y\|^2\langle z \mid x\rangle x &= \|y\|^2\langle z \mid \lambda y\rangle\lambda y \\ &= \|y\|^2\langle z \mid y\rangle\bar{\lambda}\lambda y \\ &= |\lambda|^2\|y\|^2\langle z \mid y\rangle y \\ &= \|x\|^2\langle z \mid y\rangle y\end{aligned}$$

so that the condition is also sufficient.

As for (*d*), it follows from (*b*) that $f_{x,y}$ is self-adjoint if and only if, for all $z \in V$.

$$\langle z \mid y\rangle x = \langle z \mid x\rangle y.$$

Taking $z = y$ we obtain $x = \lambda y$ where $\lambda = \langle y \mid x\rangle/\|y\|^2$. This gives $\langle x \mid y\rangle = \langle y \mid x\rangle$ whence we see that $\lambda \in \mathbb{R}$. The converse is obvious.

Test paper 1

Time allowed: 3 hours
(Allocate 20 marks for each question.)

1 Show that the subset $A = \{a_1, a_2, a_3, a_4\}$ where

$$a_1 = (1, \vartheta, 2, 1), \quad a_2 = (1, 2, 2, 1), \quad a_3 = (4, 1, 5, 4),$$
$$a_4 = (0, 3, -1, \vartheta)$$

is a basis of $\mathbb{R}^4$ whenever $\vartheta \neq 0$ and $\vartheta \neq 2$.

Suppose that $f : \mathbb{R}^4 \to \mathbb{R}^4$ is defined by $f(e_i) = a_i$ for $i = 1, 2, 3, 4$ whe $\{e_1, e_2, e_3, e_4\}$ is the standard basis of $\mathbb{R}^4$. If $\vartheta = 1$ find $x \in \mathbb{R}^4$ wi $f(x) = (-5, 5, -5, -4)$.

Show that if $\vartheta = 0$ then

$$\operatorname{Im} f = \{(a, b, c, a) \mid a, b, c \in \mathbb{R}\},$$

and that if $\vartheta = 2$ then

$$\operatorname{Im} f = \{(a, b, c, d) \in \mathbb{R}^4 \mid 16a + 3b = 7c + 8d\}.$$

2 Let $\mathbb{C}_3[X]$ be the vector space of polynomials of degree less than or eq to 2 over the field $\mathbb{C}$. Let $f \in \operatorname{Map}(\mathbb{C}_3[X], \mathbb{C}_3[X])$ be given by

$$f(1) = -1 + 2X^2,$$
$$f(1 + X) = 2 + 2X + 3X^2,$$
$$f(1 + X - X^2) = 2 + 2X + 4X^2.$$

Find the eigenvalues and the minimum polynomial of f.

3 Let V be a finite-dimensional vector space over the field F and $f \in \operatorname{Map}(V, V)$. Prove that

$$\dim V = \dim \operatorname{Im} f + \dim \operatorname{Ker} f.$$

Find a vector space V and $f \in \operatorname{Map}(V, V)$ with the property that not ev vector of V can be written as the sum of a vector in $\operatorname{Im} f$ and a vector in $\operatorname{Ker} f$.

If $f, g \in \operatorname{Map}(V, V)$ prove that

$$\operatorname{Im}(f \circ g) \subseteq \operatorname{Im} f \quad \text{and} \quad \operatorname{Ker} g \subseteq \operatorname{Ker}(f \circ g).$$

Show that $\dim \operatorname{Im}(f \circ g) \leqslant \dim \operatorname{Im} g$ and deduce that

$$\dim \operatorname{Im}(f \circ g) \leqslant \min\{\dim \operatorname{Im} g, \dim \operatorname{Im} f\}.$$

4 Prove that the product of the eigenvalues of a matrix A is $\det A$, and that if λ is an eigenvalue of an orthogonal matrix then so also is $1/\lambda$.

Deduce that if A is an orthogonal 3×3 matrix with $\det A = -1$ then -1 is an eigenvalue of A.

If B is an orthogonal 3×3 matrix with $\det B = 1$ prove that BA^{t} is orthogonal and that $\det(A + B) = 0$.

5 Find three eigenvalues and corresponding eigenvectors for the complex matrix

$$\begin{bmatrix} 1 & -\sqrt{6} & -\mathrm{i}\sqrt{2} \\ \sqrt{6} & 0 & \mathrm{i}\sqrt{3} \\ \mathrm{i}\sqrt{2} & \mathrm{i}\sqrt{3} & 2 \end{bmatrix}.$$

Hence find a matrix U such that $\bar{U}^{t}AU$ is diagonal.

Test paper 2

Time allowed: 3 hours
(Allocate 20 marks for each question.)

1 State a necessary and sufficient condition, in terms of matrix rank, fo a system of equations $Ax = b$ to have a solution.

Show that the equations

$$\begin{aligned} x + y + z + t &= 4k + 6 \\ x - y + z - t &= -2 \\ ky + 3z - (k+1)\, t &= 3 \\ 3x - 3z + (3k^2 + 2)t &= 14k \end{aligned}$$

have a unique solution except when $k = 0$ and $k = 4/3$. Show that there no solution when $k = 4/3$ and find the general solution when $k = 0$.

2 Let $\mathbb{R}_n[X]$ be the vector space of polynomials of degree less than or equ to $n - 1$ over the field $\mathbb{R}$. Define $T : \mathbb{R}_n[X] \to \mathbb{R}_n[X]$ by

$$T(f(X)) = f(X + 1).$$

Prove that T is linear and find the matrix of T relative to the basis

$$\{1, X, X^2, \ldots, X^{n-1}\}$$

of $\mathbb{R}_n[X]$. Find the eigenvalues of T. Determine the minimum polynom of T.

3 Prove that $V_1 = \{(a, b, 0) \mid a, b \in \mathbb{R}\}$ and $V_2 = \{(a, a, a) \mid a \in \mathbb{R}\}$ are subspa of $\mathbb{R}^3$. Prove also that every vector in $\mathbb{R}^3$ can be expressed uniquely as a s of a vector in V_1 and a vector in V_2.

Find $f \in \mathrm{Map}(\mathbb{R}^3, \mathbb{R}^3)$ such that

$$f \circ f = f, \quad \mathrm{Ker}\, f = V_1 \quad \text{and} \quad \mathrm{Im}\, f = V_2.$$

Show that if $g \in \text{Map}(\mathbb{R}^3, \mathbb{R}^3)$ is such that $g(V_1) \subseteq V_1$ and $g(V_2) \subseteq V_2$ then $g \circ f = f \circ g$.

4 Find an orthogonal matrix H such that $H^t AH$ is diagonal where A is the matrix.

$$\begin{bmatrix} 4\frac{1}{2} & 0 & -3\frac{1}{2} \\ 0 & -1 & 0 \\ -3\frac{1}{2} & 0 & 4\frac{1}{2} \end{bmatrix}.$$

Hence find a matrix B such that $B^3 = A$.

5 Let V be a real inner product space, let $\{e_1, e_2, \ldots, e_n\}$ be an orthonormal set of vectors in V, and let U be the subspace spanned by $\{e_1, e_2, \ldots, e_n\}$. Prove that $x \in U$ if and only if $x = \sum_{k=1}^{n} \langle x | e_k \rangle e_k$. Establish the inequality

$$(\forall x \in V) \sum_{k=1}^{n} |\langle x \mid e_k \rangle|^2 \leqslant \|x\|^2 .$$

Deduce that for $x \in V$ the following are equivalent:

(*a*) $\sum_{k=1}^{n} |\langle x \mid e_k \rangle|^2 = \|x\|^2$;

(*b*) $x \in U$;

(*c*) $(\forall y \in V) \langle x \mid y \rangle = \sum_{k=1}^{n} \langle x \mid e_k \rangle \langle e_k \mid y \rangle$.

Test paper 3

Time allowed: 3 hours
(Allocate 20 marks for each question.)

1 If $M \in \text{Mat}_{n \times n}(\mathbb{R})$ prove that, for every positive integer k,

$$I_n - M^{k+1} = (I_n - M)(I_n + M + M^2 + \ldots + M^k).$$

Deduce that if $M^{k+1} = 0$ then $I_n - M$ is non-singular.

Let

$$A = \begin{bmatrix} 2 & 2 & -1 & -1 \\ -1 & 0 & 0 & 0 \\ -1 & -1 & 1 & 0 \\ 0 & 1 & -1 & 1 \end{bmatrix}.$$

Compute $I_4 - A$, $(I_4 - A)^2$, $(I_4 - A)^3$, $(I_4 - A)^4$. By applying the first pa the question to $A = I_4 - (I_4 - A)$, prove that A is non-singular and dete mine A^{-1}.

2 Show that $\{(1, 0, 0), (1, 1, 0), (1, 1, 1)\}$ is a basis of $\mathbb{R}^3$.

A linear mapping $f : \mathbb{R}^3 \to \mathbb{R}^3$ is such that

$$f(1, 0, 0) = (0, 0, 1),$$
$$f(1, 1, 0) = (0, 1, 1),$$
$$f(1, 1, 1) = (1, 1, 1).$$

Find $f(a, b, c)$ for every $(a, b, c) \in \mathbb{R}^3$ and determine the matrix of f w respect to the basis $B = \{(1, 2, 0), (2, 1, 0), (0, 2, 1)\}$.

If $g : \mathbb{R}^3 \to \mathbb{R}^3$ is given by $g(a, b, c) = (2a, b + c, -a)$ prove that g linear and find the matrix of $f \circ g \circ f$ with respect to the basis B.

3 If r is the rank of the matrix

$$\begin{bmatrix} 1 & \alpha & 0 & 0 \\ -\beta & 1 & \beta & 0 \\ 0 & -\gamma & 1 & \gamma \\ 0 & -\delta & 1 & \delta \end{bmatrix}$$

show that

(*a*) $r > 1$;

(*b*) $r = 2$ if and only if $\alpha\beta = -1$ *and* $\gamma = \delta = 0$;

(*c*) $r = 3$ if and only if *either* $\alpha\beta = -1$ *or* $\gamma = \delta$ provided that, if $\alpha\beta = -1$, γ and δ are not both zero.

4 Let A be a real square matrix. If λ is an eigenvalue of A and $\boldsymbol{x}$ is an eigenvector associated with λ, prove that

$$\bar{\boldsymbol{x}}^{\mathrm{t}}A\boldsymbol{x} = \lambda\bar{\boldsymbol{x}}^{\mathrm{t}}\boldsymbol{x}, \quad \bar{\boldsymbol{x}}^{\mathrm{t}}A^{\mathrm{t}}\boldsymbol{x} = \bar{\lambda}\bar{\boldsymbol{x}}^{\mathrm{t}}\boldsymbol{x}, \quad \bar{\boldsymbol{x}}^{\mathrm{t}}A^{\mathrm{t}}A\boldsymbol{x} = \lambda\bar{\lambda}\bar{\boldsymbol{x}}^{\mathrm{t}}\boldsymbol{x}$$

where bars denote complex conjugates. Deduce that, if A satisfies the equation

$$A^{\mathrm{t}}A = -4I + 2A + 2A^{\mathrm{t}},$$

then every eigenvalue λ of A is such that

$$\lambda\bar{\lambda} - 2(\lambda + \bar{\lambda}) + 4 = 0,$$

and hence show that every eigenvalue of A is equal to 2.

5 Let V be the real vector space of continuous functions $f: [0, 1] \to \mathbb{R}$. Show that the prescription $\langle f | g \rangle = \int_0^1 f(t)g(t)\,dt$ defines an inner product of V. For every $f \in V$ let $T_f: [0, 1] \to \mathbb{R}$ be given by $T_f(x) = xf(x)$. Show that $T: V \to V$ described by $T(f) = T_f$ is linear and self-adjoint but has no eigenvalues.

Test paper 4

Time allowed: 3 hours
(Allocate 20 marks for each question.)

1 If F is a field and $A, B \in \mathrm{Mat}_{n\times n}(F)$ are such that $AB - I_n$ is invertibl
show that

$$(BA - I_n)[B(AB - I_n)^{-1}A - I_n] = I_n$$

and deduce that $BA - I_n$ is also invertible.

If $X, Y \in \mathrm{Mat}_{n\times n}(F)$ show that $\lambda \in F$ is an eigenvalue of XY if and on
if $XY - \lambda I_n$ is not invertible. Hence show that XY and YX have the sar
eigenvalues.

2 Let V be a finite-dimensional vector space over a field F. If A, B are subspac
of V prove that $A + B = \{a + b \mid a \in A, b \in B\}$ is a subspace of V, and th
if C is a subspace of V with $A \subseteq C$ and $B \subseteq C$ then $A + B \subseteq C$.

Let L, M, N be subspaces of V. Prove that

$$L \cap [M + (L \cap N)] = (L \cap M) + (L \cap N).$$

Give an example to show that in general $(L \cap M) + (L \cap N) \neq L \cap (M + N$

3 If S is a matrix such that $I + S$ is non-singular prove that

$$(I - S)(I + S)^{-1} = (I + S)^{-1}(I - S).$$

Deduce that $P = (I - S)(I + S)^{-1}$ is orthogonal when S is skew-symmetric.

Given that

$$S = \begin{bmatrix} 0 & \cos\vartheta & 0 \\ -\cos\vartheta & 0 & \sin\vartheta \\ 0 & -\sin\vartheta & 0 \end{bmatrix},$$

prove that

$$P = \begin{bmatrix} \sin^2 \vartheta & -\cos \vartheta & \sin \vartheta \cos \vartheta \\ \cos \vartheta & 0 & -\sin \vartheta \\ \sin \vartheta \cos \vartheta & \sin \vartheta & \cos^2 \vartheta \end{bmatrix}.$$

4 If the real matrix

$$\begin{bmatrix} a & 1 & a & 0 & 0 & 0 \\ 0 & b & 1 & b & 0 & 0 \\ 0 & 0 & c & 1 & c & 0 \\ 0 & 0 & 0 & d & 1 & d \end{bmatrix}$$

has rank r, prove that

(*a*) $r > 2$;

(*b*) $r = 3$ if and only if $a = d = 0$ and $bc = 1$;

(*c*) $r = 4$ in all other cases.

5 Let V be a real inner product space and let U be a subspace of V. Suppose that U has a basis $\{v_1, \ldots, v_n\}$. Given $x \in V$ let $x_1 = \sum_{i=1}^{n} a_i v_i$ where the coefficients $a_1, \ldots, a_n$ are given by

$$\begin{bmatrix} \langle v_1 \mid v_1 \rangle & \langle v_2 \mid v_1 \rangle & \ldots & \langle v_n \mid v_1 \rangle \\ \langle v_1 \mid v_2 \rangle & \langle v_2 \mid v_2 \rangle & \ldots & \langle v_n \mid v_2 \rangle \\ \vdots & \vdots & & \vdots \\ \langle v_1 \mid v_n \rangle & \langle v_2 \mid v_n \rangle & \ldots & \langle v_n \mid v_n \rangle \end{bmatrix} \begin{bmatrix} a_1 \\ a_2 \\ \vdots \\ a_n \end{bmatrix} = \begin{bmatrix} \langle x \mid v_1 \rangle \\ \langle x \mid v_2 \rangle \\ \vdots \\ \langle x \mid v_n \rangle \end{bmatrix}.$$

Show that $x = x_1 + x_2$ where x_2 is orthogonal to every vector in U.

In the above, call x_1 the *orthogonal projection* of x onto U and $\|x_2\|$ the *distance* of x from U.

Now let $V = C[0, 1]$ be the real inner product space of continuous functions $f: [0, 1] \to \mathbb{R}$ with inner product $\langle f \mid g \rangle = \int_0^1 f(t) g(t) \mathrm{d}t$. What is the orthogonal projection of x^2 onto the subspace $U = \langle 1, x \rangle$? What is the distance of x^2 from U?

Printed in the United States
By Bookmasters